How Real Estate Agents Can Succeed with AI

Streamline Your Business and Maximize Growth

by
Lon Welsh
Author of *The Colorado Guide to Real Estate Investing*

Joe Massey
Author of *The Mortgage Escape Plan*

About the Authors

Lon Welsh founded Your Castle Real Estate, the largest independent brokerage in Colorado, with over 700 real estate agents and 2,800+ transactions ($1.7 billion) in 2024. He co-founded First Alliance Title, a Colorado top ten title company. A former strategic consultant with Deloitte, then Arthur Andersen (Accenture) for eight years, he has been a landlord for over 20 years, peaking at over 140 residential rentals. He then sold most of the residential properties to purchase commercial buildings.

Lon has been teaching investors for over 20 years on how to thrive with real estate investing. He has written over a dozen books including *The Colorado Guide to Real Estate Investing* and has created a lot of millionaires over that time.

While helping real estate agents, their clients, and investors is Lon's passion, he has always found that true fulfillment comes from the familial bond with his team and fostering growth in the real estate community. Lon is a father of three, pursues travel, wine, golf, tennis, running, and skiing in his free time. He serves on the boards and investment committees of the Denver Zoo, the Denver/Aurora Ronald McDonald House, and Denver Boys and Girls Club.

Joe Massey is a graduate of the Colorado School of Mines with a degree in Economics. Joe has been a mortgage lender since 2002; and has been recognized by 5280 Magazine as a Five Star Mortgage Professional since 2012. Joe has been recognized as a top 500 loan originator nationwide by National Mortgage News, Scotsman Guide, Forbes and other Mortgage and Financial publications.

Joe now works as a Senior Loan Officer for Castle & Cooke Mortgage, a Billion Dollar mortgage lender based in Salt Lake City, Utah. In addition to his Loan Officer responsibilities, Joe manages several offices for Castle & Cooke Mortgage, including the Denver branch, which has been a leading office for the past 15+ years and remains a top office for Castle & Cooke Mortgage.

Joe continues to contribute to the real estate and mortgage community with his monthly presentations on financing options and continuing education. Recently he has been focused on financial education to help people gain knowledge of their mortgage so they can pay off their 30-year loan in a record time. Joe has helped hundreds of homeowners achieve the dream of owning their home free and clear through many different strategies. Joe's expertise in mortgage and personal finance has been instrumental in wealth-building and portfolio creation for many homeowners and investors.

Also by Lon Welsh

The Complete Guide to Passive Diversified Real Estate Investing, 2024
Lon Welsh
Purchase on Amazon:
https://a.co/d/7QBb7Wc

Thrive: How Real Estate Agents Can Succeed in Any Market, 2024
Lon Welsh, Rick Davidson, Eric Flohr

The Practical Guide to Colorado Real Estate Investing (2018)
The Practical Guide to Colorado Real Estate Investing (2016)
Lon Welsh, Charles Roberts, Greg Parham, Joe Massey

Game Changers, 2014
Steve Murray, Lorne Wallace, Lon Welsh

The 2013 Guide to Colorado Real Estate Investing
Lon Welsh, Charles Roberts, Jon Roberts, Tony Girard, John Dovenbarger, Greg Parham

Unlocked: Revealing the Eight Secrets of Highly Efficient Sales Professionals, 2012
Darice Johnston, Lon Welsh, Bruce Gardner, Drew Shope

The Real Estate IRA Retirement Planning Guide. 2012
Jeff Sibel, with Lon Welsh and Charles Roberts

How to Thrive with Social Media and Blogging, 2011

Also by Joe Massey

Current Editions Available Now

The Mortgage Escape Plan
How to Pay Off Your Mortgage in 7-10 Years, NOT 30!
Joe Massey
Purchase on Amazon:
https://a.co/d/i0yi6RZ

The Practical Guide to Colorado Real Estate Investing, 2021 Edition
Lon Welsh, Charles Roberts, Greg Parham, Joe Massey

The Ultimate House Hacking Guide for Denver, 2020 Edition
A Technical Guide to Building a Denver Rental Portfolio Through House Hacking
Chris Lopez, Joe Massey, Jeff White
Purchase on Amazon:
https://a.co/d/aPMKWIL

The 2022 Guide to Colorado Real Estate Investing Strategies, 2022 Edition
A Crowdsourced Book from Colorado Investors Discussing Their Strategies and Goals
Chris Lopez with contributions from Joe Massey, et. al.
Purchase on Amazon:
https://a.co/d/aOo73dn

Older Editions

The Practical Guide to Colorado Real Estate Investing, 2018 Edition
The Practical Guide to Colorado Real Estate Investing, 2016 Edition
Lon Welsh, Charles Roberts, Greg Parham, Joe Massey

Kris Andrea, Steve Depperschmidt, LaTonia Gore, Greg Parham, Lon Welsh

The 2011 Guide to Colorado Real Estate Investing
Lon Welsh, Charles Roberts, Ben Dorland, Jon Sommer, Tony Girard

Thrive: How Real Estate Agents Can Succeed in a Down Market, 2010
Lon Welsh, Bruce Gardner, Mike Welk, Drew Shope

The 2009 Guide to Denver Real Estate Investing
Lon Welsh, Charles Roberts, Michael Canon, Mike Welk

The 2021 Guide to Colorado Real Estate Investing Strategies, 2021 Edition
The 2020 Guide to Colorado Real Estate Investing Strategies, 2020 Edition
A Crowdsourced Book from Colorado Investors Discussing Their Strategies and Goals
Chris Lopez with contributions from Joe Massey, et. al.

Additional Resources From The Authors

Lon Welsh

Website: www.yourcastle.com

www.IrontonCapital.com

Linked In: www.linkedin.com/in/lonwelsh

Joe Massey

Website: https://www.masseymoneyeducation.com/payoff-your-mortgage-fast

www.masseymoneyeducation.com

www.loansbyjoemassey.com

LinkedIn: https://www.linkedin.com/in/jmasseylending/

Facebook: https://www.facebook.com/jmasseylending

Instagram: https://www.instagram.com/jmasseylending/

TikTok: https://www.tiktok.com/@jmasseylending

YouTube: https://www.youtube.com/@joemassey

Printed in the United States.

For information to reproduce selections from this book, write to:

Permissions
Ironton Capital Management, LLC
2755 S. Locust Street, #150
Denver, CO 80222

Permissions
Massey Money Education
2755 S. Locust Street #120
Denver, CO 80222

First Edition

Library of Congress Cataloguing-in-Publication Data

ISBN 978-0-9914070-4-0

Acknowledgements

"Many ideas grow better when transplanted into another mind than the one where they sprang up." — Oliver Wendell Holmes

"Alone we can do so little; together we can do so much." — Helen Keller

After over twenty years of working in real estate, I owe a debt to literally hundreds of people. I learned the most from the real estate agents at Your Castle Real Estate. I could not have figured all of this out on my own. — Lon Welsh

To Our Editors: A heartfelt thank you to our incredible editors, Serena and Kyra, for their tireless dedication, keen attention to detail, and unwavering commitment to making this book the best it can be. Your expertise, patience, and insightful feedback have truly elevated every chapter, ensuring that this book is both impactful and accessible to real estate professionals everywhere. We are beyond grateful for your hard work and this book would not be the same without you. Thank you for your invaluable contributions! — Joe Massey & Lon Welsh

Contents

OVERVIEW AND INTRODUCTION

Why We Wrote This Book

We have written many books, taught hundreds of classes, and coached many real estate agents as they have launched and grown their careers. As we have been learning about how to use Artificial Intelligence, it's become very clear that this is going to be a massive productivity enhancement for our industry. The real estate agents who embrace these tools will be capable of a vast increase in production with little additional workload compared to how hard they work today.

Who This Book Is For

This book is designed for real estate professionals who want to leverage artificial intelligence (AI) to streamline their business, enhance client interactions, and ultimately close more deals. Whether you're a brand-new agent just starting out, a seasoned real estate agent looking for a competitive edge, or a broker managing a team, this book will provide practical insights into how AI can transform your approach to real estate.

We are living in a technological revolution that is changing the way we work, and real estate is no exception. AI isn't just a buzzword; it's a set of tools that can help you automate routine tasks, analyze market trends faster, and improve customer service. By understanding how to integrate AI into your workflow, you'll not only save time but also create a more efficient and profitable business.

How To Use This Book

This book is structured to provide both foundational knowledge and actionable steps. You can read it cover to cover or jump to specific sections that are most relevant to your needs. Each chapter breaks down different aspects of the real estate business where AI can provide value. Along the way, you'll find real-world applications, tips for implementation, and best practices to ensure you get the most out of AI in your real estate career. There are many QR codes so you can download prompts, scripts, and videos. You can use these as a starting point for your custom marketing materials.

Chapter Overviews

1. AI Fundamentals

Artificial Intelligence (AI) is transforming real estate by enhancing efficiency, client engagement, and market insights. This chapter explores AI's role in automating property valuations, generating personalized client experiences, and improving marketing through predictive analytics and content creation.

Real estate agents can leverage AI tools like ChatGPT, Google's Gemini, and Microsoft's Azure AI for lead generation, market analysis, and automated responses. AI-powered chatbots, virtual assistants, and automated email tools help agents streamline communication and manage client relationships.

The chapter highlights **prompt engineering** as a critical skill for optimizing AI outputs. Writing clear, structured prompts ensure that AI delivers accurate, relevant content. Agents can refine AI-generated responses by training models in their writing style and making incremental improvements.

AI also supports **automated email generation, smart scheduling, and transaction coordination**, saving agents time and improving customer service. Unlike Google searches, AI generates tailored responses instead of retrieving existing web pages. By integrating AI strategically, real estate professionals can enhance client interactions, optimize marketing efforts, and gain a competitive edge in a rapidly evolving industry.

2. Getting Started in Real Estate with AI

AI is transforming real estate by helping agents work smarter, not harder. For new agents, AI streamlines lead generation, automates follow-ups, and provides market insights, allowing them to compete with seasoned professionals. AI-powered CRMs can handle client communication, while automated marketing tools generate property descriptions and social media content.

However, AI should enhance—not replace—personal connections. Over-reliance on automation, lack of training, and ignoring client preferences are common pitfalls. Successful agents integrate AI into their workflow to improve efficiency while focusing on relationship-building.

Top producers use AI to personalize communication, make data-driven decisions, and optimize workflows. By setting up an AI-powered CRM, refining marketing strategies, and leveraging AI for lead qualification, new agents can accelerate their success. The key is continuous learning—those who embrace AI while maintaining a human touch will dominate the future of real estate.

3. The Real Estate Agent Business Process

A structured business process is essential for real estate success, and AI can streamline each stage—from lead generation to closing. AI-powered marketing tools attract leads, while automated CRMs nurture them through personalized follow-ups. AI also enhances client consultations by analyzing preferences and providing market insights, allowing agents to tailor recommendations effectively.

For property searches and marketing, AI can generate listings, suggest targeted ads, and automate social media content. In negotiations and closings, AI-powered tools help analyze contracts, track deadlines, and ensure compliance thus reducing manual workload. AI also enables role-playing scenarios, helping agents refine their sales and negotiation skills.

While AI improves efficiency, it should complement—not replace—relationship-building. The most successful agents use AI to enhance, not automate, their client interactions. By integrating AI into daily workflows and continuously adapting, real estate agents can work smarter, provide better service, and accelerate their success in an increasingly competitive market.

4. Strategies to Attract Clients

Finding and attracting new clients is one of the biggest challenges real estate agents face, and AI is emerging as a powerful tool to make this process easier and more efficient. This chapter explores how AI can enhance lead generation, improve marketing, and streamline client communication, allowing agents to focus on building relationships and closing deals. The key takeaway is that AI doesn't replace the personal touch that makes great agents successful—it simply helps them work smarter and scale their efforts with less time and effort.

One of the most immediate ways AI can help is through **content creation**. AI tools can generate blog posts, social media captions, and even full-length video scripts, allowing agents to maintain a consistent and engaging online presence. AI-powered voice cloning and video generation make it easier than ever to produce marketing materials without spending hours recording or editing. Additionally, AI-driven social media scheduling tools help agents plan and automate posts across platforms, ensuring that their brand stays visible without constant manual effort.

Beyond content marketing, AI is revolutionizing **lead generation and qualification**. Predictive analytics tools can analyze large amounts of data to identify homeowners who are likely to sell soon, while AI-driven chatbots can interact with website visitors, answer questions, and capture new leads at any time of day. AI-generated email marketing campaigns help agents nurture leads over time, sending personalized follow-ups and reminders automatically. While these tools can't replace human intuition and relationship-building, they do ensure that agents stay top-of-mind with potential buyers and sellers without requiring constant manual outreach.

AI also enhances **client communication and relationship management** through smart CRM systems that track interactions, suggest follow-ups, and personalize messages based on client preferences. Automated referral marketing helps agents stay connected with past clients, increasing repeat business and referrals. AI-powered reputation management tools can track online reviews, suggest responses, and even generate testimonials to build trust and credibility. By keeping agents organized and proactive, AI helps ensure that no client falls through the cracks.

Finally, the chapter explores **advanced applications of AI**, such as automated event planning, market analysis, and social proof generation. AI tools can help plan client appreciation events, schedule invitations, and even analyze attendance trends to improve future outreach. Market research tools can quickly summarize pricing trends, home values, and inventory shifts, allowing agents to provide data-backed insights to clients. AI-driven social proof tools help agents showcase positive reviews, testimonials, and success stories to attract more leads.

Ultimately, the chapter emphasizes that AI is not a magic bullet, but a tool that can help real estate agents be more efficient and effective. The agents who succeed in this AI-driven era will be the ones who embrace technology while maintaining the personal relationships and trust that make this industry work. By integrating AI into their marketing, lead generation, and client management strategies, agents can scale their business, save time, and focus on what they do best—helping people buy and sell homes.

5. The Buyer Process

Buying a home is one of the most significant financial and emotional decisions a person can make. For agents, guiding clients through the buying process efficiently and effectively is critical to building trust and ensuring a smooth experience. Traditionally, agents spend countless hours explaining the same concepts—mortgage pre-qualification, closing costs, inspections, and negotiations—to each new client. With AI-powered tools and structured client education, agents can streamline this process while enhancing the buyer's understanding and confidence.

The **Buyer Video Library** is designed to serve as a scalable educational tool that empowers clients before they even step into a consultation. This 25-video series covers every major milestone, from the initial buyer onboarding to closing day and beyond. Buyers can access these videos on demand, allowing them to learn at their own pace and revisit key topics as needed. Agents can personalize these scripts to align with their brand, local market considerations, and unique client needs. By leveraging AI-generated content and automation, agents can spend less time repeating the basics and more time on high-value activities like negotiation and strategy.

A well-educated buyer is a more confident buyer. The video series covers essential topics such as mortgage pre-qualification, buyer closing costs, and understanding HOA documents, ensuring that clients are financially prepared before they start their home search. It also walks buyers through the home search and offer process, including common pitfalls, multiple-offer strategies, and contract negotiations. Once under contract, clients learn what to expect during the home inspection, appraisal, and due diligence phase, preventing surprises and ensuring smoother transactions.

Beyond the transaction, the series helps clients prepare for closing and homeownership responsibilities, from reviewing settlement statements to securing utilities and home maintenance best practices. Real estate agents can integrate these videos into their workflow by sharing them before buyer consultations, during key transaction milestones, or as post-closing educational content. By combining AI-driven automation with personalized service, real estate agents can scale their business while offering a high-touch, value-driven experience.

Author's Note: We created this entire chapter in just a few hours using AI-powered content generation. You can access the scripts online, modify them to fit your style, and even use our pre-recorded videos—or record your own. The goal is to work smarter, not harder, so you can focus on what truly matters: helping your clients find their perfect home.

6. The Seller Process

Selling a home is one of the largest financial transactions a consumer will make. Navigating the process with confidence requires knowledge, preparation, and strategic decision making. As an agent, your role extends beyond simply listing homes—you are a trusted advisor, guiding sellers through pricing strategies, market conditions, negotiations, and closing. However, educating sellers individually can be time-consuming, with many of the same questions arising repeatedly. That's where AI powered tools can help streamline the process while enhancing the seller's experience.

AI is transforming the real estate industry by providing sellers with real-time data, predictive market trends, and automated communication tools that keep them informed throughout the transaction. Instead of manually researching pricing trends, AI-driven models can generate accurate, up-to-date home valuations based on recent comparable sales, buyer demand, and local market conditions. Automated marketing platforms use AI to craft compelling listing descriptions, optimize advertising campaigns, and ensure that the right buyers see the home, increasing exposure and engagement. Additionally, AI-powered CRMs can automate follow-ups, provide personalized updates, and answer common seller questions instantly—allowing agents to focus on high-value tasks like negotiation and strategy.

This chapter explores how AI can improve every stage of the home-selling process, from setting the right price and preparing a home for showings to handling negotiations and closing the deal. A key component of this strategy is the **Seller Video Library**, a curated series of educational videos that empower sellers to understand each step of the transaction before meeting with their agent. Just like in the **Buyer Process**, these videos provide structured education, helping sellers make informed decisions while reducing repetitive conversations for agents.

By leveraging AI and video education, agents can create a more efficient, data-driven approach to home selling. This not only improves client satisfaction but also accelerates the sales process, maximizes home values, and enhances an agent's ability to scale their business. AI-powered seller education is not about replacing the personal touch—it's about making every interaction more meaningful, informed, and results-driven.

7. Role Play with AI

Success in real estate isn't just about knowing the market—it's about how well agents communicate, negotiate, and build trust with clients. Practicing these conversations is key, but finding a reliable partner to roleplay with can be difficult. This is where AI comes in. AI-powered roleplaying acts as a 24/7 practice tool, allowing agents to refine their skills in a risk-free environment. Whether it's handling objections, pitching to clients, or closing deals, AI helps agents become more confident and persuasive in real-world situations.

Even top professionals in other industries, like athletes and public speakers, continue practicing refining their skills. Real estate is no different. Agents who rehearse their sales scripts and responses to client concerns are better prepared to navigate tough conversations. AI can simulate different client personalities, objections, and negotiation scenarios, helping agents adapt their approach. From cold calling to listing presentations, AI allows agents to practice conversations as many times as needed until they feel natural and polished.

One of AI's biggest advantages is its ability to provide instant feedback and adjust to different scenarios. If a buyer hesitates, AI can simulate their concerns, prompting the agent to refine their response. If a seller objects to commission rates, AI can push back with common objections, helping the agent practice justifying their value. By engaging in these role-playing exercises, agents improve their ability to think on their feet, communicate clearly, and close deals with confidence.

Beyond structured scripts, agents can customize AI roleplaying to focus on their personal challenges. Whether it's learning how to handle multiple-offer negotiations, responding to hesitant clients, or explaining pricing strategies, AI offers unlimited opportunities for targeted training. The key to success in real estate is preparation, and AI-powered roleplaying is a game-changing tool that helps agents sharpen their skills, build confidence, and ultimately grow their business.

8. Future AI Strategies

AI is transforming the real estate industry with advanced tools that can enhance lead generation, marketing, and transaction management. While many AI applications today focus on automation and social media content development, the future will bring even more powerful tools, such as predictive analytics for identifying motivated sellers, AI-driven comparative market analysis (CMA) reports, and smart contracts that streamline transactions. These technologies will eventually help real estate agents work smarter, optimize their marketing strategies, and identify high-value leads before they even enter the market. However, while AI can provide valuable insights and efficiency, it is not a replacement for the human skills that make great agents successful—such as negotiation, relationship-building, and local market expertise.

Future AI strategies will also revolutionize marketing, branding, and client engagement. AI-powered video tools can create personalized marketing content, while advanced SEO and AI-driven social media advertising can help agents dominate local search rankings. Additionally, AI-generated investment reports and virtual reality (VR) home showings will provide buyers with more detailed insights and immersive experiences. As these AI tools continue to evolve, real estate professionals who embrace them early will gain a competitive edge, allowing them to scale their business, improve client service, and stay ahead in an increasingly tech-driven market.

9. Parting Thoughts

The real estate industry is evolving, and AI is at the forefront of this transformation. Those who embrace AI early will gain a competitive advantage, improve efficiency, and provide better service to their clients. The key is to integrate AI tools strategically, balancing automation with the human touch that remains at the heart of real estate success.

10. Glossary

Cut through the jargon with this concise AI glossary for real estate agents. Quickly understand key terms and technologies shaping the industry.

Final Words

AI is not here to replace real estate agents—it's here to **empower them**. The future of real estate belongs to those who harness AI's potential while maintaining the personal relationships that drive this industry.

1. AI FUNDAMENTALS

Introduction

Artificial Intelligence (AI) is revolutionizing the real estate industry, offering innovative solutions that enhance efficiency, accuracy, and client engagement. From automating property valuations to facilitating personalized client interactions, AI tools are becoming indispensable for modern real estate agents. This chapter delves into the fundamentals of AI, exploring its applications, benefits, and the pivotal role of prompt engineering in maximizing its potential within the real estate sector. There are plenty of examples to help you better understand how to write prompts for real estate.

What is AI?

AI is like a smart helper for real estate agents. It uses computers to think and learn, making the job of finding and selling homes easier and faster. In simple terms, AI is a tool that helps real estate agents work more efficiently, making the experience of buying or selling a home smoother for everyone.

How Can AI Help Real Estate Agents?

Artificial Intelligence is transforming the real estate industry by enhancing various aspects of a real estate agent's work. Here are some major categories where AI can assist agents, along with real estate-specific examples.

1. Property Valuation and Market Analysis

AI algorithms analyze vast amounts of data, including historical sales, market trends, and property features, to provide accurate property valuations. This assists real estate agents in setting competitive prices and advising clients effectively. *Example:* Zillow's "Zestimate" tool utilizes AI to offer real-time property value estimates, aiding both buyers and sellers in making informed decisions.

2. Personalized Client Experiences

By analyzing client preferences and behaviors, AI enables real estate agents to offer tailored property recommendations, enhancing the client experience. *Example:* Redfin employs AI to analyze user behavior on its platform, providing personalized property suggestions that match individual client criteria.

3. Automated Property Descriptions

Generative AI tools can create compelling property descriptions by analyzing features such as location, size, and amenities, saving agents time and ensuring consistency. *Example:* Some real estate agencies use AI to generate detailed property listing descriptions, highlighting key selling points to attract potential buyers.

4. Virtual Tours and 3D Imaging

AI facilitates the creation of immersive virtual property tours, allowing clients to explore homes remotely and make more informed buying decisions. *Example:* Matterport's AI-driven platform enables real estate agents to offer 3D virtual tours, enhancing the property viewing experience for buying clients.

5. Chatbots and Virtual Assistants

AI-powered chatbots can handle initial client inquiries, schedule showings, and provide property information, enhancing customer service. *Example:* Real estate firms are implementing AI chatbots to manage routine tasks, allowing agents to focus on more complex client needs.

6. Lead Generation and Qualification

AI systems can identify and prioritize potential leads by analyzing online behaviors and engagement patterns, enabling real estate agents to focus on the most promising prospects. *Example:* Revaluate uses AI to score leads based on their likelihood to move, helping agents target their marketing efforts effectively.

By integrating AI into these areas, real estate agents can enhance their efficiency, provide better client experiences, and stay competitive in the evolving real estate landscape.

Which AI Should I Use?

AI platforms have become integral tools across various industries, including real estate. Real estate agents can leverage mainstream AI platforms to enhance client services, streamline operations, and gain market insights. Here are three prominent AI platforms, along with their advantages and potential drawbacks from a real estate agent's perspective.

OpenAI's ChatGPT

Overview: ChatGPT is an AI language model developed by OpenAI, designed to generate human-like text based on user input. Real estate agents can utilize ChatGPT to automate client interactions, draft property descriptions, and provide instant responses to common inquiries.

Pros:

- **Enhanced Client Engagement:** ChatGPT can handle multiple client queries simultaneously, providing instant responses and freeing up real estate agents to focus on more complex tasks.
- **Content Generation:** The AI can assist in creating engaging property listings and marketing materials, ensuring consistency while saving time.

Cons:

- **Contextual Limitations:** While proficient in generating text, ChatGPT may occasionally produce responses that lack specific real estate context or nuance, requiring real estate agent oversight.
- **Data Privacy Concerns:** Handling client information through AI platforms necessitates stringent data privacy measures to maintain confidentiality and comply with regulations.

Google's Vertex AI

Overview: Vertex AI is Google's managed machine learning platform that enables users to deploy and maintain AI models. Real estate agents can use Vertex AI to analyze market trends, predict property values, and personalize client recommendations.

Pros:

- **Comprehensive Analytics:** Vertex AI can process large datasets to provide insights into market dynamics, helping agents and their clients make informed decisions.
- **Scalability:** The platform supports scaling AI solutions as the business grows, accommodating increasing data and user demands.

Cons:

- **Complexity:** Implementing Vertex AI may require technical expertise, potentially necessitating additional training or hiring specialists.
- **Cost Considerations:** The expenses associated with deploying and maintaining AI models on Vertex AI can be significant, especially for smaller agencies.

Microsoft's Azure AI

Overview: Azure AI is a suite of AI services offered by Microsoft, encompassing tools for machine learning, cognitive services, and bot development. Real estate agents can leverage Azure AI to enhance customer service with chatbots, analyze client sentiment, and optimize/streamline property management processes.

Pros:

- **Integration Capabilities:** Azure AI integrates seamlessly with other Microsoft products, facilitating a cohesive workflow for real estate agents using Microsoft ecosystems.
- **Robust Security:** Microsoft's commitment to security ensures that client data is protected, aligning with industry compliance standards.

Cons:

- **Resource Intensive:** Effective utilization of Azure AI may require substantial computational resources, leading to higher operational costs.
- **Learning Curve:** The breadth of services offered can be overwhelming, and real estate agents might need time to identify and implement the most relevant tools for their needs. However, incorporating these mainstream AI platforms can significantly enhance a real estate agent's capability in client engagement, market analysis, and operational efficiency.

Google's Gemini

Overview: Gemini is an AI model developed by Google DeepMind, designed to process and generate text, images, and other data types. Real estate agents can utilize Gemini to enhance client interactions, automate content creation, and gain insights into market trends.

Pros:

- **Multimodal Capabilities:** Gemini's ability to handle various data types allows agents to create comprehensive marketing materials, including property descriptions and visual content, enhancing client engagement.
- **Integration with Google Services:** Gemini integrates seamlessly with Google's suite of tools, such as Docs, Sheets, and Gmail, facilitating efficient workflow management and communication.

Cons:

- **Data Privacy Considerations:** Utilizing Gemini requires careful handling of client information to ensure compliance with data protection regulations, which may necessitate additional safeguards.
- **Subscription Costs:** Accessing advanced features of Gemini may involve subscription fees, which could be a consideration for individual real estate agents or smaller agencies.

You might want to try several options. If you extensively use Google's tools, Gemini could be a good fit since it integrates effectively with those data sources. If that isn't critical, we have found that ChatGPT, so far, to be the best option. It can also access Google data (such as files on Google Drive), but not as seamlessly as Gemini.

Understanding Prompts

What is a Prompt?

In the realm of artificial intelligence, a **prompt** serves as the initial input or instruction provided to an AI model, guiding it to generate a desired response. This input can range from a simple question to a detailed directive. Its clarity and structure significantly influence the quality and relevance of the output.

A prompt is the way humans communicate with AI systems, instructing them on the task at hand. For instance, when interacting with a language model like ChatGPT, a prompt could be a question such as, "What are the benefits of using AI in real estate?" The AI processes this input and generates a response based on its training data.

Components of an Effective Prompt

Crafting an effective prompt involves several key elements:

1. **Clarity**: The prompt should be clear and unambiguous, ensuring the AI understands the exact requirement.
2. **Context**. Providing relevant background information helps the AI generate a more accurate and contextually appropriate response.
3. **Specificity**: Clearly stating the desired outcome or format guides the AI in delivering a focused response.

For example, instead of asking, "Tell me about AI," a more effective prompt would be, "Explain how AI can assist real estate agents in property valuation."

The Role of Prompt Engineering

The practice of designing and refining prompts to elicit optimal responses from AI models is known as **prompt engineering**. This involves experimenting with different phrasings, structures, and content to guide the AI toward producing the most relevant and accurate outputs. Effective prompt engineering is crucial for leveraging the full potential of AI systems across various applications.

In summary, a prompt is the foundational input that directs an AI model's behavior. The effectiveness of AI-generated responses depends heavily on how well the prompt is crafted, making prompt engineering a vital skill for users aiming to harness AI's capabilities efficiently.

Some Sample Prompts

Here are some prompt templates designed to assist you in learning this process.

1. Identifying Potential Clients

Prompt Template:

"Analyze recent property transactions and online real estate inquiries in [specific location]. Identify individuals who have shown interest in buying or selling properties within the last [time frame]. Provide a list of potential clients along with their contact information and a brief summary of their property interests."

2. Crafting Personalized Outreach Messages

Prompt Template:

"Based on the client's interest in [property type] in [location], draft a personalized email introducing myself as a real estate agent specializing in this market. Highlight recent market trends and suggest a meeting to discuss how I can assist in achieving their real estate goals."

3. Engaging Past Clients for Referrals

Prompt Template:

"Compose an email to past clients who successfully purchased homes through my services, expressing gratitude and informing them of the current favorable market conditions. Encourage them to refer friends or family members who might be interested in buying or selling property and offer a referral incentive."

4. Local Market Snapshot

Prompt Template:

"Provide a concise overview of the current real estate market in [specific location] as of [current month and year]. Include key metrics such as median home prices, average days on market, and inventory levels. Highlight any notable trends or changes compared to the previous month."

5. Comparative Market Analysis

Prompt Template:

"Generate a comparative market analysis for [property type] in [specific neighborhood or area]. Compare the past [time frame, e.g., six months] to the previous [same time frame]. Discuss changes in average selling prices, volume of sales, and any emerging trends."

6. Investment Opportunity Overview

Prompt Template:

"Analyze the current real estate investment opportunities in [specific city or region]. Summarize factors such as rental yield trends, property appreciation rates, and economic indicators that may influence investment decisions. Provide insights into which neighborhoods are showing promising returns."

How AI is Different from a Google Search

When interacting with technology, the terms "prompt" and "search query" refer to different methods of obtaining information, each serving distinct purposes.

Search Query: A search query is a word or phrase entered into a search engine like Google to find information on the internet. The search engine matches the query with relevant web pages and provides a list of links for the user to explore. For example, typing "best restaurants in New York" into Google will yield a list of websites about top dining spots in New York City.

Prompt: A prompt is an instruction or question given to an AI model, such as ChatGPT, to generate a response. Unlike search engines that retrieve existing web pages, AI models produce new content based on the prompt. For instance, asking an AI, "Write a short story about a dragon," will result in the AI creating an original story about a dragon.

Key Differences:

- **Information Retrieval vs. Generation**: Search queries retrieve existing information from the internet, while prompts instruct AI models to generate new content or responses.
- **User Interaction**: Search engines provide a list of links for users to browse, whereas AI models deliver direct, conversational responses.
- **Content Source:** Search results are sourced from indexed web pages, while AI-generated content is produced based on patterns learned during the model's training.

Understanding these differences is crucial for effectively utilizing both search engines and AI models, as each serves unique functions in accessing and generating information. A good resource: https://learnprompting.org/ .

Train the AI in Your Writing Style

To enable an AI model to emulate your writing style, you can follow these steps:

1. Gather Writing Samples

Collect a variety of your written works that exemplify your unique style. These could include emails, articles, essays, or blog posts. The more diverse and representative the samples, the better the AI can understand your stylistic nuances.

2. Analyze Your Writing Style

Examine your writing to identify key characteristics such as tone, sentence vocabulary, and rhythm. This analysis will help in articulating your style to the AI.

3. Input Samples into the AI

Provide the AI with your writing samples. Some AI platforms allow you to input text directly, while others may require a different method. For instance, in ChatGPT, you can paste your text and instruct the AI to analyze it.

4. Create a Style Guide

Develop a style guide based on your analysis, detailing aspects like tone, preferred vocabulary, and sentence structure. This guide will serve as a reference for the AI to mimic your style.

5. Fine-Tune the AI Model

Some AI platforms offer fine-tuning capabilities, allowing you to train the model on your writing style. This process involves adjusting the AI's parameters to better replicate your unique voice.

6. Provide Feedback

After the AI generates text in your style, review it and provide feedback. Highlight areas where it aligns well with your style and areas needing improvement. This iterative process helps the AI refine its emulation of your writing.

By following these steps, you can guide an AI model to produce content that closely mirrors your personal writing style. For a visual demonstration of this process, you might find the following video helpful:

Video: Build Your Personal AI Writer (A Step-By-Step Guide) https://youtu.be/6vLlKwuzv-Y

Making Incremental Improvements

When interacting with AI models, refining your prompts through an iterative process can significantly enhance the quality and relevance of the responses. This method involves making gradual adjustments to your prompts based on the AI's outputs, thereby guiding the model toward more precise and useful answers.

Iterative Prompt Refinement

Iterative prompting is the process of systematically refining and adjusting the prompts given to a Generative AI tool to improve the relevance, accuracy, and depth of its outputs.

Example of Iterative Refinement

1. **Initial Prompt:** "Explain artificial intelligence." *AI's Response:* Provides a general overview of AI, including its definition and applications.
2. **Refined Prompt:** "Explain artificial intelligence at a 7th-grade reading level." *AI's Response:* Delivers a simplified explanation suitable for a 7th-grade student.
3. **Further Refinement:** "Explain artificial intelligence at a 7th-grade reading level, without using technical jargon." *AI's Response:* Offers an even more accessible explanation, ensuring clarity and understanding for younger audiences.

Tips for Effective Prompt Refinement

- **Be Specific:** Clearly state the desired outcome or format to guide the AI in generating a focused response.
- **Adjust Complexity:** Specify the reading level or depth of detail to match your target audience's understanding.
- **Eliminate Jargon:** Instruct the AI to avoid technical terms if simplicity is required.

By engaging in iterative prompt refinement, you can effectively guide AI models to produce content that aligns with your specific needs and audience, enhancing both clarity and relevance in the generated responses.

How AI Improves Over Time

Artificial Intelligence (AI) systems enhance their performance by learning from user interactions over time. This learning process involves analyzing user behavior, preferences, and feedback to tailor responses and improve accuracy.

Personalized Recommendations

Platforms like Netflix utilize AI to monitor users' viewing habits, such as genres and actors they prefer. By analyzing this data, the AI suggests movies and shows that align with individual tastes, enhancing user satisfaction.

E-commerce Suggestions

Amazon employs AI to track customer browsing and purchase histories. The system learns from this information to recommend products that customers are likely to be interested in, thereby increasing the likelihood of future purchases.

Music Playlists

Spotify's "Discover Weekly" feature uses AI to analyze users' listening patterns. By understanding the types of music a user enjoys, the AI curates personalized playlists, introducing users to new songs and artists that match their preferences.

Conversational AI Adaptation

AI chatbots, such as Replika, learn from ongoing user interactions. By engaging in conversations, the AI adapts to the user's communication style and topics of interest, providing more relevant and personalized responses over time

Continuous Improvement Through Feedback

AI systems often incorporate user feedback to refine their performance. For example, if a user corrects an AI's response or provides explicit feedback, the system can learn from these inputs to avoid similar errors in the future, thereby enhancing its accuracy and reliability. Through these methods, AI systems progressively align with individual user needs and preferences, delivering increasingly personalized and effective interactions.

Using AI to Take on a Specific Role for an Audience

Effectively guiding an AI model like ChatGPT to produce desired content involves clearly defining the role you want it to assume, specifying the target audience, and outlining the intended outcomes of the communication. It's a part of **prompt engineering**, which enhances the relevance and quality of the AI's responses.

Defining the AI's Role

Assigning a specific role to the AI helps tailor its responses to align with the desired perspective. For instance, instructing the AI to act as a "real estate expert specializing in neighborhood [XXX]" ensures that the information provided is relevant and authoritative within that context.

Example Prompt: "You are a real estate expert focusing on neighborhood [XXX]. Provide insights on current market trends in this area."

Identifying the Target Audience

Specifying the audience allows the AI to adjust its language, tone, and content complexity to suit the intended readers. Understanding the audience's level of awareness and familiarity with the subject is crucial.

Example Prompt: "Explain the benefits of investing in neighborhood [XXX] to first-time homebuyers with limited knowledge of the local market."

Clarifying Desired Outcomes

Clearly stating the objectives of the communication guides the AI in crafting responses that meet specific goals, whether it is to inform, persuade, or entertain.

Example Prompt: "As a real estate expert, write a persuasive article for potential investors highlighting the growth opportunities in neighborhood [XXX]."

Considering the Client's Current Level of Awareness

Assessing the client's existing knowledge about the topic enables the AI to provide information that is appropriately detailed and avoids unnecessary jargon.

Example Prompt: "As a real estate advisor, draft an introductory guide for clients new to real estate investment, focusing on the advantages of properties in neighborhood [XXX]."

By meticulously defining these elements in your prompts, you can direct the AI to generate content that is well-suited to your specific needs, ensuring clarity, relevance, and effectiveness in communication.

How to Provide Rules to Guide the Output

When crafting property descriptions for Multiple Listing Services (MLS), it is essential to adhere to Fair Housing regulations to prevent discriminatory practices. The Fair Housing Act prohibits discrimination based on race, color, religion, sex, disability, familial status, or national origin. To ensure compliance, focus solely on the property's features and amenities, avoiding any language that describes or implies preferences about potential buyers or tenants.

For instance, instead of stating "perfect for families," describe the property as "spacious home with multiple bedrooms and a large backyard." Avoid terms that could be interpreted as exclusionary or suggestive of a preferred demographic. Phrases like "ideal for young professionals" or "no children" are inappropriate and may violate Fair Housing laws. Additionally, steer clear of references to religious landmarks or institutions, such as "near St. Mary's Church," as they may imply a religious preference.

By concentrating on objective property details and steering clear of subjective descriptors, you can create inclusive listings that comply with Fair Housing guidelines.

How to Give Context and Improve Results

Providing context in AI prompts is crucial for obtaining accurate and relevant responses from language models. Context includes background information, specific details, and the desired outcome, all of which guide the AI in understanding and generating appropriate outputs. Without sufficient context, even well-structured prompts can lead to misunderstandings or irrelevant answers. For example, in professional settings, adding context to prompts makes the AI's responses more personalized, pertinent, and actionable, thereby enhancing overall communication effectiveness. By clearly defining the task, specifying the desired tone, and providing relevant details, users can improve the quality and relevance of AI-generated content.

The Importance of Context

Context serves several critical functions in communication:

- **Clarifying Meaning:** Helps recipients interpret messages accurately by providing background information.
- **Reducing Ambiguity:** Context minimizes potential misunderstandings by specifying the circumstances surrounding a message.
- **Enhancing Engagement:** Providing context makes communication more relatable and engaging for the audience. For example, when discussing market trends with potential real estate investors who lack experience, it's beneficial to explain industry-specific terms and concepts. This approach ensures that all participants have clear understanding, which facilitates informed decision-making.

Using Analogies to Simplify Complex Topics

Analogies are powerful tools for simplifying complex subjects by relating them to familiar concepts. They highlight shared characteristics between two things, making intricate ideas more accessible. For instance, explaining electrical circuits by comparing them to water flowing through pipes can help novices grasp the concept more easily.

Tailoring Communication to the Audience

Understanding your audience's background and knowledge level is crucial. When addressing professionals interested in real estate investment but lacking experience, it is important to:

- **Define Jargon:** Clearly explain industry-specific terms to avoid confusion.
- **Encourage Engagement:** Use language that motivates and instills confidence in potential investors.
- **Specify Format and Length:** Indicate if the content is intended for a blog post, video script, or another medium, and provide guidelines on length or duration. For example, instructing an AI to "Write a 500-word blog post introducing real estate investment opportunities in neighborhood X, aimed at professionals new to investing" provides clear direction for content creation.

Structuring Communication Effectively

Clear and concise communication enhances understanding. To achieve this:

- **Limit Sentence Complexity:** Avoid multiple thoughts in one sentence to prevent confusion.
- **Use Clear Language:** Steer clear of flowery or overly complex language that might obscure the message.
- **Specify Tone and Style:** Indicate whether the communication should include elements like emojis for social media or maintain a professional tone without them.

Incorporating Analogies and Perspectives

To further clarify complex topics, consider using analogies or adopting the "voice" of a well-known figure. For instance, explaining real estate investment principles in the style of Warren Buffett can provide relatable insights for the audience.

In summary, providing context in communication is vital for clarity and effectiveness. By tailoring your message to the audience, simplifying complex topics with analogies, and specifying the desired tone and style, you can enhance understanding and engagement.

Ask for Feedback

Engaging with AI models effectively involves not only crafting precise prompts but also seeking feedback on these prompts to enhance their clarity and effectiveness. By asking AI for feedback on your prompts, you can identify strengths and areas for improvement, leading to more accurate and relevant responses.

Steps to Solicit Feedback from AI on Your Prompts:

1. **Present the Original Prompt:** Begin by sharing the prompt you intend to use with the AI.
   Example: "Explain the concept of blockchain technology."
2. **Request Feedback:** Ask the AI to evaluate the prompt, focusing on aspects such as clarity, specificity, and potential ambiguities.
   Example: "Please assess the clarity and specificity of the following prompt: 'Explain the concept of blockchain technology.' What improvements can be made?"

3. **Implement Suggestions:** Based on the AI's feedback, refine your prompt to enhance its effectiveness. ***Example:*** "Explain the concept of blockchain technology in simple terms suitable for a beginner, using analogies where appropriate."

Benefits of Seeking AI Feedback:

- **Enhanced Clarity:** AI can identify vague or ambiguous language in your prompts, allowing you to make them more precise.
- **Improved Specificity:** Feedback can help you narrow down broad prompts to target specific aspects of a topic, leading to more detailed responses.
- **Identification of Assumptions:** AI can point out implicit assumptions in your prompts that may need clarification.

Example of AI Providing Prompt Feedback:

Original Prompt: "Discuss the benefits of renewable energy."

AI Feedback: "The prompt is clear but could be more specific. Consider specifying which types of renewable energy (e.g., solar, wind, hydro) and the context (e.g., environmental impact, economic advantages) to provide a more focused response."

Revised Prompt: "Discuss the environmental and economic benefits of solar and wind energy in urban areas." By engaging in this iterative process, you can fine-tune your prompts to elicit more informative and relevant responses from AI models. This approach not only improves the quality of the AI's output but also enhances your skills in effective prompt engineering.

Ask for AI to Step Back and Reconsider

Taking a step back to reflect on a problem often leads to clearer insights and more effective solutions. This approach, known as the "step-back technique," is beneficial not only for human cognition but also when interacting with AI models.

Applying the Step-Back Technique with AI

1. **Prompting for Abstraction:**

Encourage the AI to abstract from specific details to identify overarching trends or principles.
Example: "Consider the current trends in real estate investment strategies. What are the high-level patterns you observe?"

2. **High-Level Analysis Before Diving into Details:**
Ask the AI to first provide a broad overview before delving into specifics.
Example: "Provide a general overview of information overload and its impact on decision-making. Afterward, analyze how this applies to real estate investors managing extensive market data."

3. **Leveraging Training Data for In-Depth Analysis:**
Instruct the AI to utilize its training data to offer detailed insights on a particular topic.
Example: "Using your training data, provide a detailed analysis of effective strategies for converting large volumes of information into actionable insights for real estate investors."

Benefits of the Step-Back Technique

- **Enhanced Reasoning:** By prompting the AI to abstract and consider high-level concepts, it can develop a more structured and coherent response.
- **Improved Problem-Solving:** This method encourages the AI to identify fundamental principles that can be applied to specific scenarios, leading to more effective solutions.
- **Efficient Information Processing:** This technique helps in managing information overload by focusing on essential insights before exploring detailed data.

Incorporating the step-back technique in your interactions with AI can lead to more thoughtful and comprehensive responses, ultimately enhancing the decision-making process in complex fields like real estate investment.

How to Customize GPT

One of the most powerful ways to implement AI into your business is by setting up your own AI-driven GPT. Rather than relying on generic AI tools, having a custom GPT tailored to your specific business needs can significantly enhance your productivity, client interactions and marketing efforts.

Why Set Up Your Own GPT?

- **Personalized Client Engagement** – A customized GPT can be trained to reflect your tone, style, and brand, making client interactions more personalized and meaningful.
- **Lead Generation & Follow-Up** – Automate responses to inquiries, qualify leads, and schedule appointments without lifting a finger.
- **Market Analysis & Property Insights** – Instantly generate reports on property trends, home valuations, and local market conditions.
- **Content Creation** – Generate blog posts, social media content, and email campaigns with ease, ensuring consistent and professional messaging.
- **24/7 Availability** – Your AI assistant works around the clock, answering client questions and providing insights even when you're unavailable.
- **Time Efficiency** – Free up hours of your day by delegating repetitive tasks to your AI assistant, allowing you to focus on high-value activities.

<u>Steps to Setting Up Your Own GPT</u>

1. **Determine Your Objectives**
   Before setting up a GPT, define what you want it to achieve. Are you looking to automate lead responses, generate property reports, or assist with client FAQs? Clarifying your goals will help you create a focused AI assistant tailored to your business needs.

2. **Choose the Right Platform**
   Several platforms allow you to create a customized GPT, such as:
   - OpenAI's GPT Builder (ideal for creating a personalized AI assistant without extensive coding knowledge).
   - Chatbot platforms like ManyChat or Drift that integrate AI for automated conversations.
   - Custom-built solutions using GPT APIs (if you have development resources available).

3. **Train Your AI with Relevant Data**
   Your GPT is only as good as the data it learns from. Provide it with:
   - Frequently asked client questions and ideal responses.
   - Real estate industry terminology and market data.
   - Your previous email templates, blogs, and social media posts to align it with your voice.

How to Create Your Personalized GPT

For purposes of simplicity in this book and without needing any coding experience or any third-party services, we're going to focus on OpenAI's GPT Builder.

In the top right corner of Chat GPT look for the icon with your initials. Click on that icon and select "My GPTs".

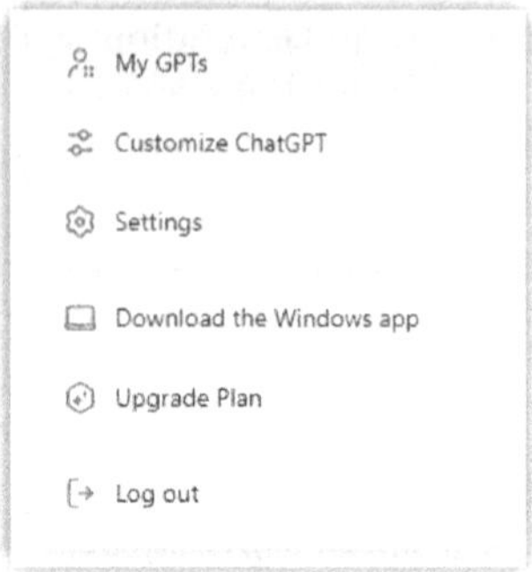

From there you will see a screen that allows you to "Create a GPT".

Click on the "+" icon to create a new GPT. It will then open a new prompt for the configuration and creation of who "you" are.

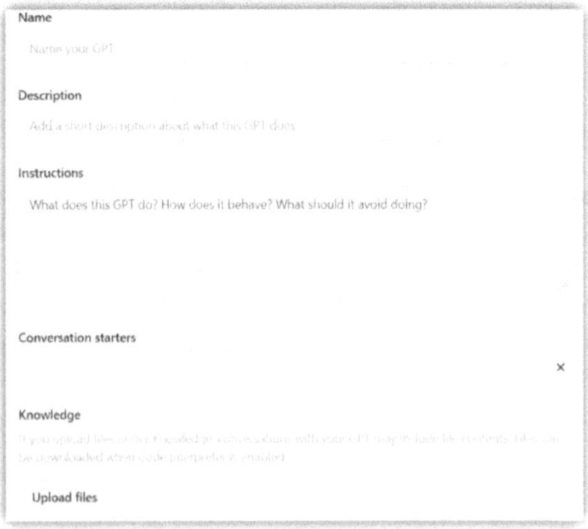

From here, you want to give it as much information as possible. For example:

- Your biography
- Your education (be specific, where you went to school, degrees earned, certifications earned)
- Your work experience (again be specific, how long in real estate, how many transactions closed)
- Reviews from former clients
- Success stories
- Links to your website
- Links to your blog
- Links to your social media (LinkedIn is great! Also Facebook, Instagram, etc.)
- Transcripts of any YouTube videos of you
- Emails to clients
- Emails to business partners
- Frequently Asked Questions that you get from your clients along with your answers to these questions
- Industry specific knowledge that you focus on – i.e. "I am an expert in {fill in the blank}"
- Specific words or phrases that you use in your daily conversations – i.e. "I don't like the industry jargon of {fill in the blank.}" or "I like to use language at a basic high school level so that all of my potential clients can understand and feel like I am communicating well."

The more information you provide to the GPT, the more it will "act" and "feel" like you and give responses similar to what you might say.

Once you have completed the setup, you will have a new GPT in your toolbar on the left. When you select this GPT, you will receive suggestions that might be relevant to your business, and the prompts that you generate will provide results in your specific style and "tone."

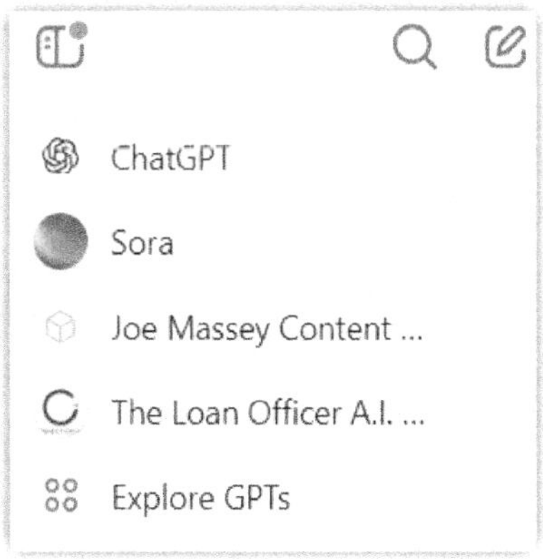

Moving forward, use this GPT when you have real estate specific needs and you will receive outputs that are very specific to your niche, your skillset, and your knowledge base.

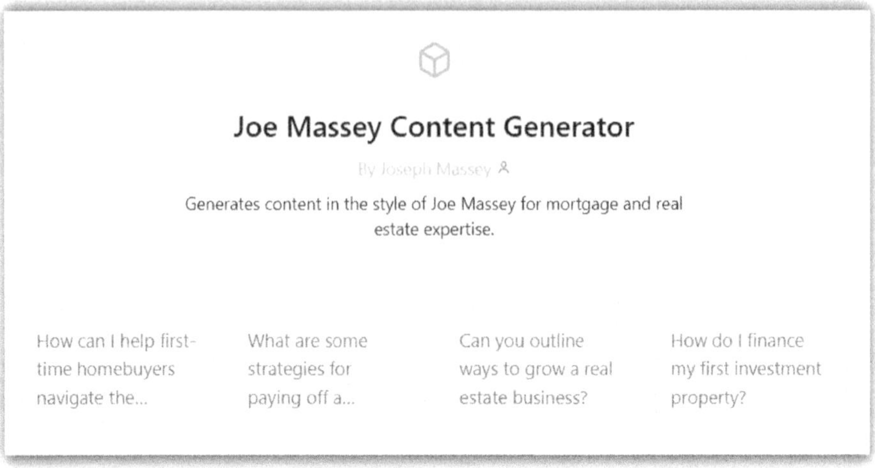

Drafting an Email

Artificial Intelligence (AI) is increasingly becoming an invaluable tool for real estate professionals, particularly in drafting effective and personalized emails. By leveraging AI, real estate agents can enhance client engagement, streamline communication, and ultimately drive business growth.

AI Applications in Email Drafting for Real Estate Agents

Personalized Client Outreach

AI tools can analyze client data to craft personalized emails that resonate with individual preferences and needs. By assessing past interactions, property interests, and demographic information, AI generates tailored messages that enhance client engagement.

Example: The real estate agent uses an AI-powered email generator to create personalized property recommendations for a client interested in waterfront properties. The AI crafts an email highlighting new listings that match the client's preferences, increasing the likelihood of a successful sale.

Automated Follow-Up Emails

Timely follow-ups are crucial in real estate transactions. AI can automate this process by scheduling and drafting follow-up emails based on client interactions and timelines, ensuring consistent communication without manual effort.

Example: After a property viewing, an AI system schedules a follow-up email to the potential buyer, expressing appreciation for their visit and providing additional information about the property, thereby keeping the client engaged.

Market Updates and Newsletters

Keeping clients informed about market trends is essential. AI can generate informative newsletters and market updates by analyzing current real estate data, ensuring clients receive relevant and up-to-date information.

Example: The real estate agent employs AI to compile and draft a monthly newsletter that includes recent market trends, new listings, and investment opportunities, providing value to clients and establishing the real estate agent as a knowledgeable resource.

Case Study: AI-Powered Email Implementation

A real estate agency integrated an AI email generator into their client communication strategy. The AI analyzed client data to personalize outreach, resulting in a 30% increase in email open rates and a 25% boost in client engagement. Additionally, automating follow-up emails reduced the administrative burden on agents, allowing them to focus more on client interactions and closing deals.

Crafting Effective Prompts for AI Email Generation

To maximize the benefits of AI in drafting emails, agents should provide clear and detailed prompts. Consider the following structure:

- **Role Definition:** Specify the role you are assuming.
  Example: "As a Real estate agent specializing in luxury properties..."

- **Purpose of the Email:** Clarify the objective.
  Example: "...draft an email to inform clients about a new beachfront listing..."

- **Desired Emotional Response:** Indicate the emotion you wish to evoke.

Example: ...that evokes excitement and urgency."

- **Format Specification:** Refine the format and length.
  Example: "...in a concise format suitable for a quick read."

Comprehensive Prompt Example:
"As a real estate agent specializing in upscale neighborhoods, draft an email to inform potential buyers about a new listing near top-rated schools. The email should evoke a sense of community and warmth, encouraging families to envision their future in this home. Please keep the email under 200 words."

Best Practices for AI-Generated Emails

- **Review and Edit:** Always review AI-generated emails to ensure accuracy and appropriateness.
- **Maintain Authenticity:** Ensure the email reflects your personal voice and brand.
- **Compliance:** Verify that the content adheres to industry regulations and company policies.

By integrating AI into email drafting, Real estate agents can enhance their communication strategy, foster stronger client relationships, and improve overall efficiency in their operations.

Chapter 1 Summary

In this chapter, we have explored the foundational aspects of AI and its transformative impact on real estate practices. We began by defining AI and discussing its relevance to real estate agents, highlighting how it aids in tasks such as market analysis, property valuation, and client communication. The chapter emphasized the importance of understanding and crafting effective prompts to guide AI tools, ensuring outputs are accurate and aligned with specific objectives. By mastering prompt engineering, real estate agents can harness AI to streamline operations, enhance decision-making, and provide personalized services to clients, thereby staying competitive in an increasingly digital marketplace.

2. GETTING STARTED WITH AI

If you're just getting started in real estate, you are stepping into an exciting and competitive industry. The good news? You have access to tools that previous generations of agents could only dream of —AI is your secret weapon. Whether you're looking to find more clients, save time, or make smarter decisions, AI can help you work faster and more effectively than ever before. But the key is knowing how to use it wisely. AI won't replace the relationships and personal touch that make real estate successful, but it will help you become a more productive, knowledgeable, and confident agent.

One of the biggest advantages AI offers new agents is the ability to quickly analyze and understand market trends. Instead of spending hours sifting through MLS data, AI-powered tools can highlight patterns, predict which neighborhoods are heating up, and even suggest pricing strategies for listings. This means that even if you are new to real estate, you can make data-driven decisions with the confidence of a seasoned pro. AI can also help you identify potential buyers and sellers before your competition does by analyzing online behavior, search trends, and financial indicators. In short, AI gives you an edge in knowing where the opportunities are.

Beyond market insights, AI makes client communication easier and more personalized. Imagine having an assistant that can draft follow-up emails, schedule appointments, and even remind you when it's time to check in with a lead—all without you lifting a finger.

AI-powered chatbots can answer basic client questions, giving you more time to focus on high-value activities like negotiations and showings. The key is to use these tools to enhance, not replace, your personal connections with clients. Real estate is still a relationship business, and AI should be your behind-the-scenes partner, making sure nothing falls through the cracks while you build trust with your clients.

Finally, one of the smartest ways to start using AI in real estate is to automate repetitive tasks. Managing social media posts, generating property descriptions, and even creating targeted ads can all be handled by AI, freeing you up to do what you do best—helping people buy and sell homes. By integrating AI into your workflow from day one, you'll avoid common rookie mistakes, stay organized, and position yourself as a forward-thinking agent who delivers results. The future of real estate belongs to those who embrace technology while keeping the human touch at the center of their business. AI can't replace you, but it can make you an unstoppable force in the industry.

For new real estate agents, starting a real estate business can feel overwhelming. AI can help simplify the learning curve by automating tasks and providing valuable insights. Here's how AI can assist early in your career:

- **Lead Generation Made Easy:** AI can analyze online behavior, social media activity, and property search patterns to help you identify potential buyers and sellers before your competition does.

- **Smart Market Insights:** Instead of spending hours researching market trends, AI tools can quickly provide data on pricing, neighborhood trends, and demand, helping you make informed decisions.

- **Automated Client Follow-Ups:** AI-powered CRMs can schedule follow-up emails, texts, and reminders, ensuring that no lead slips through the cracks while you focus on building relationships.

- **Effortless Listing Descriptions and Marketing:** AI can generate compelling property descriptions, automate social media posts, and create targeted ad campaigns, saving you time and boosting your visibility.

- **Time-Saving Administrative Support:** From organizing contracts and documents to setting up appointments and reminders, AI-powered tools help you stay organized and focused on revenue-generating activities.

Let's start by understanding the trajectory of agents from Newbie to Top Producer.

Agent Career Trajectory

Typical agent career trajectory from Newbie to High Producer.

ILLUSTRATIVE

	New Agent	Emerging Agent	Scaling Up	Master Agent
Learning Focus	Learning "craft" skills • MLS, CMA • Listing presentations • Writing contracts	Learning marketing skills • Ninja • Buffini • Proctor	Mastering marketing skills • Lead generation • SOI management • Gaining efficiencies	Learn (master) time mgmt • Time blocking • How to say no, etc. • Getting leverage
Client Attributes	Few clients • Work with anyone • Lots of bad clients • Scarcity mentality	More clients • Work with *almost* anyone • Very bad clients screened out; still tolerate most • Often still scarcity view	Enough clients • Becoming selective • Bad clients screened out; "here's how I work" • Abundance mentality	Too many clients • Agent choses not to work with just *any* client • Mostly referral driven • Clients appreciate you
Process & Support	• What's a process? • Post it notes and stack of business cards (often not called back)	• Emerging use of CRM ("is Excel OK?") • Fumbling use of tx coord • Rough processes	• Good to strong CRM • Strong processes • Often strong use tx coord • Fumbling use of admin	• Excellent CRM • Excellent processes • Good to excellent use of assistants and small team
Results	• Discounted commissions • Limited, no income • Often lots of stress	• Struggle to get prospecting done consistently • Income roller coaster, stress (working with bad clients, unstable income, not enough income)	• Good income • Working lots of hours (nights, weekends) • Family frustration • Cause of stress shifts from income to time	• Very good income • Hours are reduced • Protect family time and personal "me" time

Source: YCRE analysis

Across, the top of the chart are four skill levels that most real estate agents will progress through, on their journey from New to Master Agent. On the left side, you can see four skill tracks:

- Learning Focus
- Client Attributes
- Process and Support
- Results

Pick a skill track and read across from left to right to get a sense of how your skills should build over time.

Learning Focus

As a New Agent, you need to learn craft skills. AI will be immensely useful in helping you roleplay. How to ask for a referral, how to ask for an appointment, how to explain and defend your commission – these are all excellent applications for AI. See the Buyer Process and Seller Process chapters for an in-depth exploration.

The current state of AI doesn't help with CMA (home listing pricing), but we imagine that will improve over time.

As an Emerging Agent, many of the marketing tasks can be partially to totally automated with AI. See the chapter on Strategies to Attract Clients to learn more.

As a real estate agent that is Scaling Up, you are going to get more effective with your CRM (customer relationship manager) and consistent follow-up. There might be some AI applications, but not as many as for Newer and Emerging Agents.

Finally, as a Master, time blocking, time management, and leverage will be your learning focus areas. AI currently has a rudimentary level of skill to replace an assistant or VA (virtual assistant, usually offshore, at a lower price). We expect this will be an area of intense focus for AI development firms. The more VA roles an AI can take on, the more the AI companies can charge us for their monthly subscriptions. Watch for developments here.

How AI Can Help Agents with Learning Focus at Each Stage

- **New Agent: Mastering the Basics with AI Roleplay**
    - AI-powered chatbots and virtual assistants can help new agents practice essential skills such as asking for referrals, setting appointments, and handling commission objections. By simulating real client conversations, AI allows agents to build confidence and refine their scripts before engaging with actual leads.

- **Emerging Agent: Automating and Enhancing Marketing Efforts**
    - At this stage, AI can assist with automating email campaigns, creating social media content, and optimizing digital ads. AI tools can analyze engagement data to suggest the best marketing strategies, helping agents become more efficient in attracting and nurturing leads.

- **Scaling Up to Master Agent: Leveraging AI for Time Management and CRM Optimization**
    - As agents grow their business, AI-powered CRMs can streamline client follow-ups, prioritize high-value prospects, and automate scheduling. At the Master level, AI-driven virtual assistants can help with time blocking, managing administrative tasks, and ensuring agents stay focused on high-impact activities, allowing them to scale effectively.

Client Attributes

As a New Agent, you won't have enough clients. That will lead most agents to "cling" to every prospect, even if it is not a good client. Prospects that won't get prequalified for a loan, won't sign a buyer agency, or lack motivation are typical poor-quality clients. Master Agents won't work with them. AI won't help you here. What AI will help with is your marketing, so you have more prospects. That will give you the confidence to say "no" (and say it faster).

Emerging Agents and Agents Scaling Up get progressively better about identifying and screening out poor prospects. Having more leads is really what tends to move agents down the learning curve.

How AI Can Help Improve Your Client Attributes

- **Attract More Qualified Leads with AI-Driven Marketing**
  - AI can analyze online behavior, search trends, and demographic data to target high-quality leads who are more likely to convert. By using AI-powered ad campaigns and automated follow-ups, new agents can generate a steady flow of prospects, reducing the urge to cling to unqualified clients.

- **Automate Lead Scoring to Identify Serious Buyers and Sellers**
  - AI-powered CRMs can analyze engagement levels, prequalification status, and past interactions to score leads based on their likelihood to transact. This helps Emerging and Scaling Agents prioritize motivated clients while filtering out tire-kickers and timewasters.

- **Personalize Communication to Build Trust Faster**
  - AI tools can tailor emails, texts, and social media outreach based on client interests and behaviors. By delivering the right message at the right time, agents can nurture relationships more effectively and increase their chances of working with serious, motivated clients.

Process and Support

Newer Agents usually don't have any processes. AI can make it easier for you to learn and implement the processes that differentiate the Master from the New Agent. You should be able to get down the learning curve much faster as a result. By far, the most important process is getting and consistently using a CRM. The more rapidly you master this, the better. Then it's a challenge to grow your network (or database, or SOI/sphere of influence – all mean the same thing) to the correct size. There's a linear relationship between the size of your SOI and your income. AI assistance in marketing will help you grow your SOI and then stay in touch with them consistently.

How AI Can Help with Process and Support

- **New Agent: Implement a CRM and Automate Follow-Ups**
  - AI-powered CRMs help new agents establish structure by automating client follow-ups, organizing contacts, and scheduling reminders. This ensures consistent communication and prevents leads from slipping through the cracks, helping agents build a strong foundation early on.

- **Emerging Agent: Use AI to Expand and Engage Your SOI (Sphere of Influence)**
  - AI-driven marketing tools can help agents grow their database by identifying potential leads through social media activity, website visits, and online behavior. AI can also personalize outreach campaigns to keep prospects engaged, strengthening relationships over time.

- **Scaling Up to Master Agent: Optimize Efficiency with AI-Powered Workflows**
  - As agents grow their business, AI can streamline transaction management, automate administrative tasks, and assist with scheduling. AI-powered virtual assistants can handle routine processes, allowing agents to focus on high-value activities like negotiations, client meetings, and closing deals.

Top 5 Ways Rookie Agents Can Use AI to Speed Up Their Trajectory

As a new real estate agent, you're juggling a lot—finding clients, marketing properties, learning market trends, and handling endless paperwork. It's easy to feel overwhelmed, but AI can be a game-changer, helping you work smarter instead of harder. By automating repetitive tasks, providing deep market insights, and enhancing client communication, AI allows rookie agents to compete with seasoned pros right from the start. Whether it's generating leads, crafting high-quality marketing materials, or streamlining follow-ups, AI helps you focus on what really matters—building relationships and closing deals. Here are five powerful ways AI can help you hit the ground running in your real estate career.

1. Find and Attract More Clients

AI-powered lead generation tools can analyze online behavior, social media activity, and search patterns to help new agents identify motivated buyers and sellers. Instead of relying on cold calling or door-knocking, AI helps you focus on high-quality leads who are more likely to convert into clients.

2. Create Professional Marketing

AI can generate eye-catching property descriptions, automate social media posts, and even design marketing materials like flyers and ads. With AI-driven tools, you don't need to be a marketing expert—just provide the basics, and AI will create polished, engaging content that attracts attention.

3. Automate Client Communication and Follow-Ups

New agents often struggle to keep up with lead follow-ups, but AI-powered CRMs can handle this for you. AI chatbots and automated messaging systems can send follow-up emails, text reminders, and appointment confirmations, ensuring you stay top-of-mind with clients without spending hours on manual outreach.

4. Get Instant Market Insights and Pricing Strategies

AI-driven analytics tools can quickly analyze MLS data, market trends, and recent sales to help you determine the best pricing for a listing or spot undervalued properties for buyers. This means you can provide expert-level insights, even if you're new to the industry.

5. Save Time on Administrative Tasks

AI can streamline paperwork, manage contracts, and organize documents, reducing the time spent on tedious administrative work. Virtual assistants powered by AI can also help schedule showings, set reminders, and handle repetitive tasks—giving you more time to focus on closing deals and building relationships.

As you can see, AI can be a powerful tool. However, you need to be careful to become too dependent on it. The bulk of the real estate industry has always been (and likely will always be) based on personal connections. Utilizing AI can help you facilitate more conversations and increase your speed from Rookie to Master Agent, but make sure you don't get "lost in the weeds" of AI and not spend enough time on the personal connections. In the following section we address some of the common mistakes that we see rookies make in their initial training and use of AI.

Common Mistakes New Agents Make When Using AI

Over-Relying on Automation

AI can handle follow-ups, generate marketing content, and streamline workflows, but it can't replace genuine human connection. Clients want to feel heard and valued, not just processed through an automated system. Agents who rely too heavily on AI risk losing the personal touch that builds trust and long-term relationships. The best approach is to let AI handle repetitive tasks while you focus on meaningful, client-centered interactions.

Lack of Training

AI tools are powerful, but they're only as effective as the person using them. Many new agents adopt AI without fully understanding its capabilities, leading to underutilization or even costly mistakes. Learning how to properly use AI for lead generation, marketing, and client follow-ups will ensure you maximize its benefits without falling into the trap of automation overload. Investing time in training early on will set you up for long-term success.

Ignoring Client Preferences

AI can generate emails, social media posts, and even property recommendations, but generic, one-size-fits-all messaging won't resonate with clients. Agents who blindly use AI-generated content without tailoring it to individual clients' needs risk alienating potential buyers and sellers. The key is to use AI as a tool for personalization—leveraging its insights to craft messages that feel customized and relevant, not robotic.

Failing to Integrate AI into Existing Workflows

Many agents try AI tools as standalone solutions without fully integrating them into their CRM, email systems, or marketing platforms. This creates inefficiencies, making it harder to track leads, follow-ups, and transaction progress. AI works best when it's part of a larger, well-structured workflow, seamlessly connecting different aspects of your business rather than operating in isolation.

Overestimating AI Capabilities

While AI is great at analyzing data, predicting trends, and automating processes, it can't replace the human skills required in real estate. Negotiations, relationship-building, and understanding client emotions are areas where human expertise is irreplaceable. Agents who expect AI to handle everything—from pricing a home to closing a transaction—will be disappointed. The smartest agents use AI as an assistant, not as a replacement for the essential human elements of the business.

How Future Top Producers Use AI to Succeed

Hyper-Personalize Communication

Top-producing agents know that one-size-fits-all messaging doesn't work in real estate. AI allows them to tailor emails, text messages, and marketing campaigns based on a client's behavior, preferences, and search history. For example, AI can analyze a buyer's browsing patterns on a real estate website and automatically send them listings that match their style and budget. Similarly, AI-driven CRMs can track when a past client is likely to move again and trigger a personalized follow-up message. By using AI to send the right message at the right time, top agents make their communication feel natural, relevant, and engaging; leading to stronger relationships and higher conversion rates.

Focus on Data-Driven Decisions

Rather than relying on gut instinct, top agents use AI-powered insights to make smarter business decisions. AI can analyze MLS data, market trends, and recent sales to help agents determine optimal pricing strategies for listings. It can also assess social media engagement and ad performance to refine marketing campaigns, ensuring that money is spent where it delivers the highest return. For neighborhood analysis, AI tools can predict which areas are becoming hot markets, helping agents position themselves ahead of competitors. By letting AI handle the heavy lifting on data analysis, top producers make well-informed, strategic decisions that maximize their success.

Enhance Client Relationships

AI isn't just about automation; it's about making clients feel valued. Top agents use AI-powered CRMs to track client preferences, milestones, and past interactions, allowing them to provide personalized service at scale. AI can remind an agent when a past client's home purchase anniversary is approaching, prompting a thoughtful check-in message. AI can also anticipate needs based on behavioral patterns. If a buyer is suddenly active on listing websites again, the agent can be notified to follow up with relevant options. By leveraging AI to stay proactive, top producers make their clients feel cared for, leading to more referrals and repeat business.

Master Tech Integration

The best agents don't just use AI—they ensure it works seamlessly within their existing systems. They integrate AI tools with their CRM, MLS, email marketing platforms, and social media accounts to create a fully connected workflow. For example, AI-powered chatbots can instantly sync lead information into a CRM, which then triggers automated follow-ups based on lead behavior. AI-driven scheduling tools can coordinate showings and client meetings without manual back-and-forth. By ensuring all AI tools communicate with each other, top producers reduce administrative burdens and create a more efficient, scalable business.

Continuously Learn and Adapt

AI is constantly evolving, and top agents stay ahead by keeping up with the latest advancements. They attend webinars, take courses, and experiment with new AI-driven tools to refine their strategies. They also track how AI is being used in other industries, adapting successful tactics for real estate. For instance, AI-driven sentiment analysis—which is used in e-commerce—can be applied to real estate to gauge a client's emotional response to listings. By staying curious and open to innovation, top producers ensure they never fall behind, making AI an ongoing competitive advantage in their business.

Action Steps to Go from Rookie to Top Producer with AI

If you're a new agent looking to fast-track your success, AI can be a powerful but only if you use it strategically. The agents who become top producers don't just adopt AI; they integrate it into their daily workflows, allowing them to work smarter, not harder. Your goal is to build habits and systems that set you up for long-term success. Here's exactly what you should be doing right now to move from rookie to top producer using AI.

Step 1: Set Up Your AI-Powered CRM and Automate Your Follow-Ups

Your CRM is the backbone of your business. Start by choosing a CRM that offers AI-powered automation for lead follow-ups, email campaigns, and client reminders. Set up automated responses for new inquiries so no lead ever slips through the cracks. AI can also help you score leads based on engagement, so you know who to prioritize. Make it a daily habit to review your CRM and follow up with top prospects—your consistency here will pay off in closed deals.

Step 2: Use AI to Generate and Refine Your Marketing Strategy

AI can take the guesswork out of marketing. Use AI-driven tools to create social media posts, ad campaigns, and email content that resonate with your audience.

Instead of blindly spending on ads, leverage AI analytics to track performance and optimize your budget. For content creation, AI can generate listing descriptions, blog posts, and even video scripts, saving you hours each week. Focus on personalizing these AI-generated materials to maintain an authentic connection with your audience.

Step 3: Implement AI for Lead Generation and Qualification

Don't waste time chasing bad leads. AI-powered chatbots and lead capture forms can help you pre-qualify prospects by gathering essential information before you even speak to them. AI-driven ad platforms can target potential buyers and sellers based on their online behavior, ensuring you reach the right audience. Make it a weekly goal to analyze which lead sources are working best and adjust your strategy accordingly.

Step 4: Integrate AI with Your Daily Operations for Maximum Efficiency

AI works best when it is part of a seamless workflow. Make sure your AI tools integrate with your CRM, email system, MLS, and social media platforms. Use AI scheduling tools to automate appointment booking and AI transaction management tools to keep deals on track. The more you automate routine tasks, the more time you'll have to focus on high-value activities like negotiations and client relationships.

Step 5: Stay Curious, Keep Learning, and Adapt

AI is evolving fast, and the agents who stay ahead of the curve will dominate the industry. Commit to ongoing education—whether it's watching AI tutorials, attending real estate tech webinars, or experimenting with new AI tools. Test different AI applications in your business and refine what works best. Stay connected with top-producing agents who are successfully using AI, and don't be afraid to pivot your strategy as technology advances.

By taking these actionable steps, you'll not only streamline your business but also set yourself up for long-term success. The key is to balance automation with a strong personal touch—using AI as your assistant while you focus on building relationships and closing deals. The agents who embrace AI today will be the top producers of tomorrow. Will you be one of them?

Chapter 2 Summary

AI is a powerful tool for new real estate agents, helping them navigate the competitive industry by automating tasks, offering market insights, and enhancing client communication. AI tools can assist with lead generation, providing data-driven insights on pricing and trends, automating client follow-ups, and streamlining administrative duties like scheduling and paperwork. This technology enables rookie agents to compete with seasoned professionals by saving time and boosting productivity. However, it's crucial to balance AI's automation with the personal touch that is essential for building strong client relationships. While AI can handle repetitive tasks and improve efficiency, it can't replace human skills such as negotiation and relationship-building, which are central to real estate success. For new agents, integrating AI effectively into their workflows, continuously learning, and maintaining a focus on personal connections will set them on the path to becoming top producers.

3. REAL ESTATE AGENT BUSINESS PROCESSES

Why Every Real Estate Agent Needs a Business Plan

A real estate career without a business plan is like driving to an unfamiliar destination without a GPS—you might eventually get there, but you'll take longer, waste time, and miss valuable opportunities along the way. A well-structured business plan provides clarity, helping you set clear goals, define your target market, and outline actionable strategies to generate leads, close deals, and grow your income. It also ensures you allocate your time and resources efficiently, focusing on high-impact activities rather than reacting to whatever comes your way.

Beyond guiding your daily activities, a business plan helps you track progress and make adjustments as needed. The real estate market is always evolving, and agents who have a plan can adapt to changes more effectively than those who operate without direction. AI can enhance this process by providing data-driven insights, helping you analyze market trends, optimize marketing campaigns, and improve lead conversion strategies.

Successful real estate agents don't just work hard, they work smart. A business plan helps you stay focused, organized, and intentional about building a thriving real estate career. Whether you're a new agent or looking to scale your business, taking the time to create and refine your business plan will give you a roadmap to long-term success.

How AI Can Help with Your Business Plan

AI can be a powerful tool in developing a real estate business plan by providing data-driven insights and automating key components of the planning process. Instead of starting from scratch, agents can use AI-powered templates and chat-based tools to generate customized business plans based on their goals, target market, and experience level. AI can analyze local market trends, predict potential lead sources, and even suggest marketing strategies based on successful industry practices. Additionally, AI-driven CRMs can track progress on business goals, offering real-time adjustments and recommendations to improve efficiency. By leveraging AI, real estate agents can create a dynamic, data-backed business plan that evolves with market conditions, helping them stay competitive and focused on the right strategies for long-term success.

Let's see what AI might come up with if we ask it to help with a business plan.

Sample 1

I am a rookie real estate agent. I want you to act as a real estate coach with 25 years experience helping new agents become top producers. Please give me a structured business plan with action steps that I should follow. My goal is to make $100,000 within the next 12 months.

And AI gives us the following:

12-Month Real Estate Business Plan to Earn $100,000

This structured plan is designed to help you reach your goal of earning $100,000 in your first year as a real estate agent. You'll focus on lead generation, branding, networking, conversion, and efficiency.

1. Set Your Income Goals and Break Down the Numbers

- **Target Income:** $100,000
- **Average Commission Per Sale:** Estimate ~$5,000 per transaction (depends on your market)
- **Total Transactions Needed:** 20 homes closed ($100,000 ÷ $5,000)
- **Leads to Closings Ratio:** If 10% of your leads convert, you need 200 strong leads.
- **Monthly Goal:** Close 2 transactions per month.

2. Build Your Sphere of Influence (Weeks 1-4)

- **Create a List of 200+ Contacts**
  - Friends, family, colleagues, social media contacts, past clients from previous careers.
  - Use a CRM (e.g., KVCore, Follow Up Boss, LionDesk) to track interactions.
- **Announce Your Career on Social Media & in Emails**
  - Share your story: Why did you become an agent? How can you help?
  - Post about your journey and market trends weekly.

3. Lead Generation: 4 Key Sources (Ongoing)

A. Open Houses (Weekly)

- Host at least **2 open houses per weekend** (8 per month).
- Follow up with every visitor via phone, text, and email within 24 hours.

B. Cold Calling & FSBO/Expired Listings (Daily)

- **Call 25-50 homeowners per day.** Use a script and set appointments.
- Focus on:
    - FSBOs (For Sale By Owners)
    - Expired Listings
    - Just Listed / Just Sold Calls

C. Online Lead Generation

- **Facebook & Instagram Ads**: Spend **$500–$1,000 per month** on lead ads.
- **Google PPC Ads**: Target buyers and sellers searching for realtors.
- **Create Video Content**: 2-3 videos per week on real estate topics.

D. Networking & Referrals

- **Attend 1-2 networking events per month** (chamber of commerce, investor meetups).
- **Ask for referrals from past clients, friends, and family** regularly.

4. Master Your Follow-Up System

- Phone Calls & Texts:
  - **Follow up with new leads within 5 minutes.**
  - Call and text them at least **5 times in 10 days** until they respond.
- Email Campaigns:
  - Weekly market updates, home-buying tips, and testimonials.
- In-Person Appointments:
  - Aim for at least **5 in-person meetings per week**.

5. Become a Local Market Expert

- Daily Market Research (15 Min/Day)
  - Study **MLS hot sheets** for new listings, price reductions, and market trends.
- Weekly Content Creation (1-2 Hours)
  - Share market insights via video, email, and social media posts.

6. Improve Sales Skills

- Role-play **scripts daily** (handling objections, presenting to sellers).
- **Read/listen to 1 sales book per month** (e.g., "The Millionaire Real Estate Agent").
- Shadow top producers in your office.

7. Time Management

Task	Time Block
Lead Generation & Prospecting	9:00 AM – 11:00 AM
Follow-Ups	11:00 AM – 12:00 PM
Appointments & Showings	1:00 PM – 4:00 PM
Content Creation & Branding	4:00 PM – 5:00 PM
Admin & Training	5:00 PM – 6:00 PM

8. Scaling Up & Getting to $100K Faster

- By month **6**, reinvest profits into paid leads or an **ISA (Inside Sales Agent)**.
- Build relationships with **real estate investors** for repeat business.
- Consider adding **a niche** (luxury homes, first-time buyers, relocations).

9. Track Your Progress & Adjust

- **Weekly Reviews:**
  - How many calls did you make?
  - How many appointments were set?
  - How many closings?
- **Monthly Check-Ins:**
  - Adjust strategies based on conversion rates.

Final Words

- **Commit to daily prospecting & follow-up.**
- **Build relationships, not just transactions.**
- **Stay consistent – success comes from disciplined execution.**

Would you like help setting up a lead generation system or scripts for calls and follow-ups?

Wow! In less than one minute, AI has given us a complete business plan to show us what we should be doing, how we should be doing it and exactly what activities we should be focused on. And it even offers suggestions to help us setup a lead generation system and scripts for our calls and follow-ups. So, it really IS a real estate coach with 25 years' experience!

Agent Business Process

Now we have a clear and concise business plan. Let's see how AI can help us with our Business Process.

A successful real estate agent follows a structured process from lead generation to closing deals. AI enhances each stage:

- Lead Generation
- Lead Nurturing and Follow-up
- Client Consultation and Needs Analysis
- Property Search and Marketing
- Negotiation and Closing

Lead Generation

The first step in a thriving real estate business is finding potential buyers and sellers. AI-powered marketing tools can help agents attract leads through targeted ads, social media insights, and predictive analytics. Automated chatbots and AI-driven CRM systems can also capture and qualify leads, ensuring agents focus on the most promising prospects.

How Can You Use AI in Your Lead Generation?

A great example on how to start with AI in generating leads would be to have AI help with your social media to promote an open house. Here's a sample prompt that you can use:

I am a real estate agent who is doing an open house at "123 Main Street." This property is in Denver, Colorado. It is a 3 bedroom, 2 bathroom house with a 2 car garage. It has mountain views on a quiet street. Can you write me 3 social media posts that I can use to advertise this property?

Please give me the following for the social media posts:
1 - Title
2 - Video Script
3 - Caption with emojis

Please output this in a table format. I want you to write as a real estate expert with 25+ years of experience in marketing open houses. Make the posts fun, engaging and likely to get interaction from neighbors and potential homebuyers.

Let's see what AI came up with:

Social Media Open House Posts

Title	Video Script	Caption with Emojis
🏡 Your Dream Home Awaits! 🏠	🎥 *[Standing in front of the house]* "Hey everyone, it's [Your Name] with [Your Brokerage], and I am standing in front of the most charming 3-bedroom, 2-bathroom home right here in Denver! 🌟 This home has **breathtaking mountain views**, a **quiet street** for peace and relaxation, and a **spacious 2-car garage** for all your storage needs! 🏡✨ *[Walks toward the front door]* You'll love the **open floor plan**, the **modern kitchen**, and the **cozy backyard**—perfect for summer BBQs! 🍔🔥 Don't miss our **Open House this weekend**—come see for yourself! *[Camera pans to the mountain view]* Can you imagine sipping your morning coffee with this view? ☕⛰️ Comment 'DREAM HOME' below if you'd love to check it out!"	🚨 **OPEN HOUSE ALERT!** 🚨 This **3-bed, 2-bath** Denver home has **MOUNTAIN VIEWS** and sits on a **peaceful street!** 🏡🏠 Don't miss your chance to see it—join us this weekend for a tour! 🎉 Tag a friend who's house hunting! 👇👇 #DreamHome #DenverHomes #OpenHouse
😍 Mountain Views + Cozy Vibes!	🎥 *[Inside the living room, showing the view through a window]* "Step inside 123 Main Street and prepare to fall in love! ❤️ This **bright, open-concept** home is perfect for entertaining and cozy nights in. 🌟 *[Walks toward the kitchen]* Check out this **chef's kitchen**—tons of counter space, stainless steel appliances, and natural light pouring in! ✨ *[Pans to the backyard]* And the backyard? It's calling your name for summer BBQs and morning coffee sessions! ☀️☕ Come see it in person—join us for an **Open House this weekend!** 📅 Will we see you there? Drop a 'YES' in the comments!"	🔥 **Open House This Weekend!** 🏡 ✨ Come tour this stunning **3-bed, 2-bath** home with **mountain views**, an **open floor plan**, and a **peaceful backyard!** 🏠🏡 Who do you know that needs to see this? 👀 Tag them below! 👇👇 #DenverRealEstate #OpenHouseVibes #DreamHome
🏡 Love Where You Live!	🎥 *[Walking up the driveway]* "They say home is where the heart is... and this home? It's got **ALL the heart!** ❤️ Nestled on a quiet street with **mountain views**, this **3-bedroom gem** has everything you need! ✨ *[Walks inside, shows the living area]* A spacious, sunlit living room? Check! ✅ *[Moves to the kitchen]* A modern kitchen ready for your culinary creations? ✅ *[Walks outside]* And a **perfect backyard oasis** for relaxing and entertaining? ✅✅ Join us this weekend to see this beauty in person! Drop a 🏡 in the comments if you're coming by!"	📢 **Stop scrolling!** This **Denver home** has everything you're looking for—**mountain views**, a **quiet street**, and an **open, inviting floor plan!** 😍 Join us for the **Open House this weekend!** 🎉 Who's coming? Drop a 🏡 below! #HouseGoals #DenverLiving #MountainViews

In less than a minute we have three social media posts that we can use. All we need to do is upload the video script to a teleprompter app, hit the record button, read what we see on the screen and post to our social media channels. Pretty powerful way to start on your lead generation? We'd say so!

Lead Nurturing and Follow-Up

Not every lead is ready to buy or sell immediately. AI-driven CRMs can automate follow-ups, sending personalized emails and reminders based on client behavior. AI chatbots can also keep leads engaged by answering common questions and providing relevant property listings in real time.

But what if syncing up your AI with your CRM is too advanced for you at this stage? Is there anything else you could do with AI to help improve your lead nurturing and follow up campaigns?

Let's stick with the open house theme and see how AI can help us with another sample prompt.

> I'd like you to write me a 10 step campaign to follow up with people that I met at the open house for 123 Main Street. Please give me a combination of phone calls, text messages and emails that I can send to these potential prospects.
>
> Please output in table format:
> Column A - Date and time to follow up
> Column B - Follow up method (phone, text, email)
> Column C - Script I should use
>
> Please write as a real estate expert with 25 years experience in doing open houses. I want to sound motivational and positive about real estate and helping them find a new home. I want to be engaging, but not pushy or "salesey"

What did AI come up with?

10-Step Open House Follow-Up Campaign

Date & Time	Method	Script
Day 1 (Same Day, 5:00 PM - 7:00 PM)	📱 Text Message	"Hi [First Name], this is [Your Name] from [Your Brokerage]! It was great meeting you today at the open house at **123 Main Street** 🏡✨ What did you think? I'd love to hear your thoughts! Let me know if you have any questions—I'm happy to help! 😊"
Day 2 (Morning, 9:30 AM - 11:30 AM)	📞 Phone Call	"Hi [First Name], it's [Your Name] from [Your Brokerage]! Just wanted to check in and see what you thought about **123 Main Street**. I know sometimes things hit differently after a night to sleep on it! 😊 Are there any questions I can answer for you? Or would you like to see similar homes in the area? I'd love to help!"
Day 3 (Evening, 6:00 PM - 7:30 PM)	📧 Email	**Subject:** Thinking About 123 Main Street? 🏡 "Hey [First Name], I hope you're having a great day! I wanted to send over a quick **recap of the home you toured** at **123 Main Street**—along with a couple of similar homes in the area that might interest you. 🏡✨ If you'd like to take a second look or explore more options, let me know! I'd love to help you find the perfect home. Let's chat soon! Best, [Your Name]" (Include a couple of links to similar homes.)
Day 5 (Afternoon, 1:00 PM - 3:00 PM)	📱 Text Message	"Hi [First Name]! Just wanted to check in—are you still thinking about **123 Main Street**? 🏡🏡 It's such a great home in a fantastic neighborhood! If you're interested, I can show you **similar homes** that fit your style and budget. Let me know if you'd like to see a few! 😊"
Day 7 (Morning, 10:00 AM - 12:00 PM)	📞 Phone Call	"Hey [First Name], I wanted to follow up to see if you're still in the market for a new home! Whether it's **123 Main Street** or another home that's a better fit, I'd love to help you navigate your options. What's been on your mind about your home search? Let's chat!"
Day 10 (Evening, 6:30 PM - 8:00 PM)	📧 Email	**Subject:** More Homes Like 123 Main Street? "Hi [First Name], I know that finding the perfect home takes time! Since you showed interest in **123 Main Street**, I wanted to send over some similar homes that might be a great fit for you. 🏡💡 Let me know if any of these catch your eye—I'd be happy to schedule a private tour! Best, [Your Name]" (Include a few more listings.)
Day 14 (Midday, 12:00 PM - 2:00 PM)	📱 Text Message	"Hey [First Name], I just came across a **new listing** that I think you'd love! 😍 Would you like me to send over the details? Also, if you want to revisit **123 Main Street**, I can set that up for you too! Let me know what you think! 🏡😊"
Day 21 (Morning, 10:00 AM - 12:00 PM)	📞 Phone Call	"Hi [First Name], I wanted to touch base and see how your home search is going! Have you seen anything that caught your eye? I'm always here if you need advice, listings, or private tours. Let me know how I can help!"
Day 30 (Evening, 6:00 PM - 8:00 PM)	📧 Email	**Subject:** Still Looking for the Right Home? "Hey [First Name], I wanted to check in and see where you're at with your home search. 🏡 If you're still looking, I'd love to refine your search and send over homes that truly match your wishlist! Let me know what's most important to you, and I'll find the best options. Hope to hear from you soon! 😊 Best, [Your Name]"
Day 45 (Afternoon, 1:00 PM - 3:00 PM)	📱 Text Message	"Hi [First Name]! Just thinking of you and your home search! 🏡 Have you found the right place yet, or would you like me to keep sending listings that match your needs? No pressure at all—just here to help however I can! 😊"

In less than a minute of work, AI has given us a complete follow-up campaign that we can load into our CRM. When we meet a prospect at the open house, all we need to do is put their information into the CRM. The CRM will give us an action item at every step of the way and tell us exactly what to say in our follow-ups. Powerful stuff and we're just scratching the surface.

Client Consultation and Needs Analysis

Once the lead is ready, agents must understand their client's needs, preferences, and financial situation. AI can assist by analyzing past interactions, predicting client preferences, and providing market data to help agents make data-driven recommendations.

This is where AI can be instrumental in helping us hone our skills and our role-playing to make sure we have a great client appointment. What if we worked with AI to plan and practice the appointment:

> Now, I want to practice a script and role play for how I would interact with this client during a client appointment. The scenario is:
>
> The prospect has been receiving our follow-up messages. They have now responded that they are open to meeting to discuss their needs and goals for buying a new home. Please write me a 3-5 minute script of how this meeting should go.
>
> I'd like you to include conversational points that I should start with, along with expected questions that I'll get from the client. Include answers to these questions that I should have prepared.

What ideas does AI produce?

1. Warm Welcome & Setting the Tone (First 30-60 seconds)

[With a warm, confident tone]

"Hey [First Name], great to meet with you today! I really appreciate you taking the time to chat about your home search. I know buying a home is a big step, and my goal today is to really understand **what you're looking for, what's most important to you, and how I can make the process as smooth and stress-free as possible.**

We'll cover a few key things today—like your must-haves, your timeline, and what the process looks like—but most importantly, I want to answer any questions you have. Sound good?"

(Pause for agreement—this builds rapport and gets them engaged early.)

---

2. Discovery Questions – Understanding Their Needs (2-3 Minutes)

[Start with broad questions, then narrow down based on their answers.]

1. **"Tell me, what has you thinking about buying a home right now?"**
(Listen for motivations like upsizing, downsizing, investment, job relocation, etc.)

2. **"What does your dream home look like? What's a 'must-have' vs. a 'nice-to-have'?"**
(Encourage them to list out their priorities—location, size, layout, style, yard, etc.)

3. **"Have you already been pre-approved for a mortgage, or would you like recommendations for a great lender?"**
(Gauge their financial preparedness—if not pre-approved, you can connect them with a lender.)

4. **"Are there specific neighborhoods or areas you're drawn to?"**
(Understand their preferences and whether they are open to other suggestions.)

5. **"Are there any deal-breakers? Things you absolutely don't want?"**
(This helps filter out homes that won't be a good fit.)

3. Addressing Common Buyer Questions (2-3 Minutes)

🔅 *[Anticipate and be prepared for these common concerns.]*

💬 **Buyer Question 1: "How does the home-buying process work?"**

◆ *Your Response:*

"Great question! The process has a few key steps:

1️⃣ **Pre-approval** – This helps us know what price range to focus on.

2️⃣ **Home search** – We'll tour homes and narrow down options.

3️⃣ **Making an offer** – Once we find the perfect place, I'll guide you through making a strong offer.

4️⃣ **Inspection & appraisal** – This ensures the home is a good investment.

5️⃣ **Closing day!** – Once everything checks out, you get the keys to your new home! 🎉"

💬 **Buyer Question 2: "How competitive is the market right now?"**

◆ *Your Response:*

"Right now in [Denver], the market is [briefly summarize: hot, balanced, or buyer-friendly]. What that means for you is [explain if homes are selling fast, if there's room to negotiate, etc.]. I'll make sure you have **the right strategy** to find the best home and get a great deal!"

💬 **Buyer Question 3: "Do I need 20% down to buy a home?"**

◆ *Your Response:*

"Actually, no! Many buyers put down **as little as 3-5%**, depending on the loan type. There are also **down payment assistance programs** available. I can connect you with a lender who can walk you through all your options!"

💬 **Buyer Question 4: "How do I know I'm getting a good deal?"**

◆ *Your Response:*

"That's where I come in! I analyze the market daily, so I'll make sure you're not overpaying. I also run a **comparative market analysis** to ensure the home is priced fairly before we make an offer. My job is to help you find the right home at the right price!"

---

4. Next Steps & Closing the Meeting (Last 1 Minute)

📌 *[Recap the key points and set expectations for the next steps.]*

"Based on everything we've talked about, I have a really clear picture of what you're looking for, and I already have a few homes in mind that might be a great fit. Here's what we'll do next:

✅ I'll send you a list of homes that match your criteria within the next [timeframe].

✅ Let's set up a few home tours—are you available [suggest dates]?

✅ If you haven't gotten pre-approved yet, I can introduce you to a trusted lender to get started.

Does that sound good? I'm excited to help you find your new home!"

Final Thought: Make It Personal!

🌟 *End on an upbeat, motivational note:*

"Buying a home is such an exciting journey, and I'm here to **make it as smooth and enjoyable as possible**. I'll guide you every step of the way so you never feel overwhelmed. I can't wait to find the perfect home for you!"

Outcome:

By the end of this meeting, the buyer should feel **excited, informed, and confident** about working with you. You've built trust, answered their key concerns, and set the next steps in motion!

If you're a new agent and you're nervous about meeting with potential buyers or you don't know what to say, now you have a script you can practice in the mirror or even role-play with AI to know what to say. (See the Role-Playing section later in this chapter.)

Property Search and Marketing

For buyers, AI can curate personalized property recommendations based on search history and preferences. For sellers, AI-powered tools can generate professional listing descriptions, create targeted marketing campaigns, and analyze market trends to optimize pricing strategies.

Following our same clients from the open house, let's see if AI can help us put together some ideas for where we might want to show homes.

> Now I want you to help me plan for where I should take these buyers to show them homes. Their target needs are:
>
> 3 bedrooms
> 2 bathrooms
> Back yard if possible, but also open to townhomes with shared outdoor space
>
> They both work in the Denver Tech Center. They want to be no more than a 20 minute drive to work. They enjoy the outdoors, they have a dog that they like to walk and they want to be near any potential open spaces or parks. What areas would you recommend that we begin for our search?

What results does AI deliver?

Based on your clients' preferences and requirements, here are several neighborhoods within a 20-minute drive to the Denver Tech Center (DTC) that offer 3-bedroom, 2-bathroom homes, access to parks and open spaces, and are dog-friendly:

1. **Greenwood Village**

- **Proximity to DTC:** Located within the Denver Tech Center, offering a minimal commute.
- **Outdoor Amenities:** Home to numerous parks and trails, including the High Line Canal Trail and Cherry Creek State Park, providing ample space for outdoor activities and dog walking.
- **Housing Options:** Features a mix of single-family homes and townhomes, many with spacious yards or access to shared green spaces.

2. **Centennial**

- **Proximity to DTC:** Approximately a 10-15 minute drive to the Denver Tech Center.
- **Outdoor Amenities:** Boasts over 80 miles of trails and numerous parks, such as DeKoevend Park and Willow Creek Park, ideal for dog owners.
- **Housing Options:** Offers a variety of housing styles, including single-family homes with backyards and townhomes with communal areas.

3. **Highlands Ranch**

- **Proximity to DTC:** Roughly a 20-minute drive to the Denver Tech Center.
- **Outdoor Amenities:** Features extensive open spaces, trails, and parks, including Chatfield State Park, which is great for outdoor enthusiasts and dog owners.
- **Housing Options:** Predominantly single-family homes with generous yard spaces.

*Edited for brevity: AI actually gave 10 different recommendations.

Are you starting to see the power of AI and how it can speed up your client interactions? You may be reading this and saying "Yeah, I already know all of this…" But think about how you can use this to create concise descriptions of the different areas. You could share this with your clients and set up MLS searches in the various areas to quickly narrow down which areas they prefer or dislike as you begin the home search.

Negotiation and Closing

AI-powered contract analysis tools can streamline paperwork, identify potential issues, and ensure compliance. AI can also assist in tracking transaction timelines and automating reminders for critical deadlines, ensuring a smooth closing process for both agents and clients.

Here's an example of how you can use AI to streamline this process.

Sample Contract.pdf
PDF

Great news, I have this client under contract. I am uploading the contract here. Can you please analyze the dates in the contract and give me Outlook Calendar appointments that I can download and save for all of the crucial dates in the contract? Especially important are the resolution dates (inspection, appraisal, loan termination and of course closing.

And what does AI do for us?

I've created an Outlook-compatible calendar file with all the crucial contract dates, including inspection, appraisal, loan availability, and closing deadlines.

You can **download the file here** and import it into Outlook:

Download Open_House_Contract_Dates.csv

Let me know if you need any modifications!

Ok, kind of cool that AI can read the contract and give me the dates, but it's a CSV file…

Subject	Start Date	Start Time	End Date	End Time	All Day Event	Description	Location	Private
Inspection Termination Deadline	2/15/2025	9:00 AM	2/15/2025	10:00 AM	TRUE	Reminder for Inspection Termination Deadline	N/A	TRUE
Inspection Objection Deadline	2/15/2025	9:00 AM	2/15/2025	10:00 AM	TRUE	Reminder for Inspection Objection Deadline	N/A	TRUE
Inspection Resolution Deadline	2/18/2025	9:00 AM	2/18/2025	10:00 AM	TRUE	Reminder for Inspection Resolution Deadline	N/A	TRUE
Appraisal Deadline	2/17/2025	9:00 AM	2/17/2025	10:00 AM	TRUE	Reminder for Appraisal Deadline	N/A	TRUE
Appraisal Objection Deadline	2/18/2025	9:00 AM	2/18/2025	10:00 AM	TRUE	Reminder for Appraisal Objection Deadline	N/A	TRUE
Appraisal Resolution Deadline	2/19/2025	9:00 AM	2/19/2025	10:00 AM	TRUE	Reminder for Appraisal Resolution Deadline	N/A	TRUE
Loan Availability Deadline	2/21/2025	9:00 AM	2/21/2025	10:00 AM	TRUE	Reminder for Loan Availability Deadline	N/A	TRUE
Closing Date	2/28/2025	9:00 AM	2/28/2025	10:00 AM	TRUE	Reminder for Closing Date	N/A	TRUE
Possession Date	3/3/2025	9:00 AM	3/3/2025	10:00 AM	TRUE	Reminder for Possession Date	N/A	TRUE

Seems like a lot of work to copy this over to Outlook. I don't think that really saved us any time. Can AI help make it easier?

> Can you walk me through how to import this to outlook?

> Yes! Here's a step-by-step guide on how to **import the Outlook calendar file** I created into Microsoft Outlook:
>
> **Step-by-Step Guide to Import the Calendar File into Outlook**

Author Note: The authors chose not to include the step-by-step guide as this is not a "training manual" but rather a guide on creativity and how you can improve your processes and save time with AI.

The resulting output from AI was a six-step process that included precisely what to click on, which buttons to toggle and exactly the steps to have the contract calendar integrated with your Outlook calendar so you can track the dates and deadlines. In less than two minutes and with step-by-step instructions, the contract is completely integrated into your daily calendar, and you don't have to try to track these dates on a notepad or (even worse) in your head. Talk about time management!

Other Buyer Process Ideas

Negotiation Role Play

When researching this book, we talked to several buyer agents that role-played with AI on a negotiation point with the listing agent. You can take the role play prompt and adjust it. For example, your prompt could be:

> *Imagine you are an experienced Buyer Agent. You are about to go under contract on a house with solar panels. This is the last remaining negotiation point before you go under contract. The solar panels are leased from the builder, and the Buyer doesn't want to take over the lease. The Buyer is insistent that the Seller pay off the lease before closing.*
>
> *You are role-playing with a tough Seller's Agent. Their client, the Seller, is short on cash. They feel like they came down on the price and they have "given enough" for the deal, and do not want to give this concession as well.*

Either role play the dialogue back and forth for me or let me play the role of the Buyer's Agent, and you (the AI) play the role of the Listing Agent.

We'll cover Role Playing in more detail in Chapter 8.

Document Summaries

Several of the real estate agents we interviewed for the book use AI to develop summaries of title policies or HOA documents. In one example, the HOA package ran to over 500 pages. We asked the AI to generate a 5-10 page summary of the most important things to know from the package. It selected quite a few things that seemed relevant. You'll want to coach your client that these summaries are NOT a replacement for their own detailed review of the documents or seeking legal advice. It's just a tool to make the process easier.

One best practice with that 500+ package might be to ask the AI several smaller questions.

- *Please summarize the key points from the recent association meeting minutes. Provide the page number in the PDF if I need to look at the detailed text for more information on any of your summary points.*

- *Please summarize the key points from the covenants and restrictions. Provide the page number in the PDF if I need to look at the detailed text for more information.*

Avoid over-reliance on automated systems. Balancing AI insights with human judgment is key to ethical and effective negotiation practices.

Using AI to Streamline and Elevate Your Business

The modern real estate business is built on efficiency, consistency, and client service—all areas where AI can be a game-changer. As we've seen throughout this section, AI is not just about automation; it's about helping you work smarter at every stage of the real estate process. From generating leads with social media prompts to streamlining follow-up campaigns and integrating contracts into your calendar, AI enhances your ability to serve clients and stay organized. The key takeaway? The more you embrace AI, the faster you'll build the habits and systems that separate top producers from struggling agents.

However, AI is not a magic bullet—it's a tool that amplifies the work you put in. Simply having AI generate scripts or automate emails won't make you successful unless you apply what AI provides in a meaningful way. You still need to build relationships, refine your negotiation skills, and deliver excellent service. The agents who win in this AI-driven era will be the ones who combine technology with their personal touch, using AI to enhance rather than replace human connection.

So, where do you go from here? Start by integrating AI into just one part of your process—perhaps by automating your follow-ups or using AI to refine your client consultations. Once you're comfortable, expand AI into your marketing, lead generation, and transaction management. The goal isn't to overhaul everything overnight, but to progressively improve your efficiency and effectiveness.

Chapter 3 Summary

A business plan is essential for real estate agents as it provides direction, clarity, and a structured approach to achieving goals. It helps agents define their target market, generate leads, close deals, and grow their income efficiently. AI can play a significant role in developing and refining these plans by offering data-driven insights, automating tasks, and enhancing decision-making. It helps agents with lead generation, follow-up campaigns, client consultations, and even contract management. Real estate agents can gradually integrate AI into their workflows, starting with specific areas like follow-ups and client consultations, to improve efficiency and effectiveness over time. As you can see, AI streamlines processes, improves client interactions, and provides valuable market analysis, but it must complement human judgment and relationship-building.

4. STRATEGIES TO ATTRACT CLIENTS

Welcome to the most important chapter in the book. We've coached real estate agents for 20+ years, and the biggest struggle for many is getting enough clients. Learning how to show homes and write contracts is a skill that most can develop a lot faster than prospecting.

In our brokerage, we talk extensively about relationship selling. AI is not going to change that. This is an intensely personal business, and that is why AI is not going to replace real estate agents. Keep your primary focus on maintaining those great relationships. AI will be the tool to help you do it with less effort.

The big gain for you from AI is when you want to acquire new clients. These tools, used well, will really accelerate the growth of your business.

In this chapter, we explore how real estate agents can leverage AI to enhance client acquisition and prospecting efforts. The focus is on practical applications of AI tools to streamline processes, improve communication, and effectively target potential clients.

Here's what we'll discuss:

- AI-Powered Content Creation
- AI-Driven Lead Generation and Qualification
- Enhancing Client Communication and Relationship Management
- SEO Optimization and Market Analysis
- Optimizing Your Online Presence, Social Proof and Reputation Management
- Innovative Applications of AI

We'll also offer our assessment on whether each idea is available now, may be soon, or is AI hype.

<u>Priority</u>

V High	Applies to essentially all agents
High	Applies to most agents
Med	Applies to agents with a special marketing niche
Low	Most agents can skip today

Status

Now+	Available now and one of the first to be adopted; You might be behind
Now	Available now with emerging adoption; You can lead
Emerging	Available, but not fully baked

Hype

Reality	Works well and ready for everyone
Mostly Reality	Works OK; might be better for early adopters
Midpoint	Best for agents that really need it; most should wait
Hype	Not really out of beta

Effort

Easy

Medium

Hard

Here's how to prioritize your efforts:

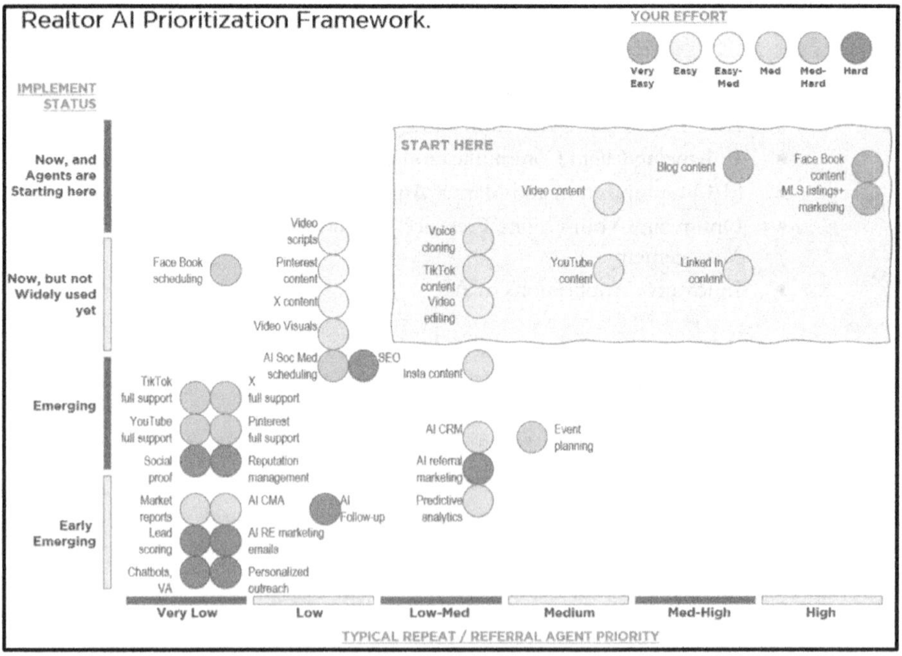

The rankings and prioritization are from the point of view of a typical production agent that:

- Gets most of their leads from repeat and referral.
- Doesn't have a lot of desire to increase their social media (SM) presence to get more leads.
- They would like to be more effective with SM to engage their current database.
- They want to be current with AI, but they don't want to be bleeding edge.

We will cover each of these ideas on the following pages.

AI-Powered Content Creation

Voice Cloning for Content Creation

Status:	**Available Now**
Your priority:	**Low to Medium**
Hype v. Reality:	**Mostly Reality**
Effort:	**Low**
Examples:	https://elevenlabs.io/ use your voice
	https://play.ht/ use your voice
	https://speechify.com/ use a celebrity voice
	https://www.wellsaid.io/ professional readers

Voice cloning is a technology that allows you to create a digital copy of your voice using AI. This means you can generate speech that sounds just like you without having to record every word. For real estate agents, this can be a powerful tool to personalize communications and save time.

Author Note: This worked really well for Joe. For Lon, it works well for shorter (five-minute blogs) pieces, but didn't work well for creating an entire Audible version of a book. There's a bit of one-time effort to get this set up; then it's pretty simple once you have it configured.

How Does Voice Cloning Work?

Voice cloning uses AI to analyze recordings of your voice. The AI learns the unique patterns, tones, and nuances that make your voice distinct. Once trained, the AI can produce new speech in your voice, allowing you to input text and receive audio that sounds like you are speaking.

Steps to Clone Your Voice:

1. **Choose a Voice Cloning Tool:** Several platforms offer voice cloning services. For example, ElevenLabs provides instant voice cloning with just a few minutes of audio, and it is the platform the authors have used the most. Another option is PlayHT, which offers high-quality voice cloning suitable for various applications.
2. **Record Your Voice:** To train the AI, you'll need to provide audio samples of your voice. This can be done by uploading existing recordings or by reading a provided script. Clear audio with minimal background noise will yield the best results.
3. **Upload and Train:** Once your recordings are ready, upload them to the chosen platform. The AI will analyze your voice to create a digital clone. This process can vary in duration depending on the tool used.
4. **Generate Speech:** After training, you can input text into the platform, and it will generate audio that sounds like you are reading the text. This is useful for creating personalized messages, videos, virtual tours, or automated responses.

Benefits for Real Estate Agents:

- **Personalized Outreach:** Enhance client relationships by sending personalized voice messages to clients without recording each one individually.
- **Rapid Social Media:** Create informational videos more easily to attract new clients.
- **Consistent Branding:** Reinforce your personal brand by maintaining a consistent voice across all communications.
- **Time Efficiency:** Quickly generate voice content for listings, updates, and newsletters, saving valuable time.

Considerations:

While voice cloning offers many advantages, it's essential to use it responsibly. Always inform clients when using AI-generated voices to maintain trust. Additionally, ensure that the use of voice cloning complies with local regulations and ethical guidelines. Incorporating voice cloning into your real estate practice can modernize your communication strategy, making interactions more personal and efficient.

Blog Content Generation

Status: **Available Now+**

Your priority: **Medium**

Hype v. Reality: **Reality**

Effort: **Easy**
Examples: Any AI tool that you like, such as Gemini, ChatGPT, Claude
https://write.homes/ AI for real estate marketing

Creating engaging blog content is essential for real estate agents to attract and inform clients. Artificial Intelligence (AI) can simplify this process by generating relevant and appealing content.

Author Note: Of the agents we interviewed for this book, this was one of the first things they used AI for. After the development of MLS and marketing descriptions, it seems to be one of the first things agents try. This might be a good place for you to start. It's pretty easy, and it's a great way to learn how to write great prompts.

How Does It Work?

AI tools analyze vast amounts of data to understand language patterns and topics. By inputting specific prompts or topics, these tools can generate coherent and contextually relevant blog posts tailored to your needs. For instance, platforms like ChatGPT can produce content based on the information you provide.

Steps to Do It:

1. **Choose an AI Tool:** Select a user-friendly AI content generator suitable for real estate topics. Tools like Write. Homes offer various templates for emails, blogs, and social media posts.
2. **Input Your Topic:** Enter the subject you want to write about. For example, "Tips for First-Time Homebuyers in Denver." This was the starting point for the chapters in this book on the Buyer Process and Seller Process.
3. **Generate the Content:** Let the AI create a draft based on your input.
4. **Customize the Content:** Specify any particular points or style preferences. Some tools allow you to set the tone or include specific keywords. For example, our Buyer Process video on Inspections is generic to geography. If you work in an area where well and septic systems are common, you would want to customize that – it will really show off your expertise.
5. **Review and Edit:** Read through the generated content to ensure it aligns with your voice and provides accurate information.

Benefits for Real Estate Agents:

- **Time Efficiency:** AI can quickly produce drafts, allowing you to focus on other tasks.
- **Consistency:** Maintains a steady flow of content, keeping your audience engaged.
- **SEO Optimization:** Some AI tools incorporate relevant keywords to improve search engine rankings.

- **Time Saving:** By leveraging AI for blog content creation, real estate agents can maintain an informative and engaging online presence with reduced effort.

Considerations:

- **Accuracy:** Always verify the information provided by AI to ensure it is correct and up to date.
- **Reasonableness**: A problem with AI that is not being addressed is how reasonable the output is. For example, if you ask AI how it can help real estate agents, it will say it can predict who the next client will be, to save you time. As we are about to see, that is technically accurate – there are some products that do that – but they don't work that well. They are something you have to buy from another company; not something you can do on your own.
- **Personal Touch:** While AI generates content, adding your personal insights can make the blog more relatable.
- **Cost:** Some advanced AI tools may require a subscription, so consider your budget.

Social Media Content Generation – Facebook

Status:	**Available Now**
Your priority:	**High to Very High** (for just content creation) **Low** (for scheduling and all other add-ons)
Hype v. Reality:	**Reality**
Effort:	**Easy** (for just content creation) **Hard** (for scheduling and all other add-ons like monitor engagement)
Examples:	Any AI tool that you like, such as Gemini, ChatGPT, Claude www.write.homes AI for real estate marketing www.RealEstateContent.ai

Author Note: Most agents rely on repeat and referral for their business, and Facebook is an important cornerstone to stay in touch with their database. Content generation might be an initial place for you to start. That's easy. Integrating all these ideas (e.g., scheduling) will be a lot more set-up and work, but once you have it ready, it will save a lot of labor going forward.

How Does It Work?

Platforms like www.buffer.com, www.hootsuite.com, and www.canva.com provide AI-powered assistance for content scheduling, graphics, and caption generation. All the same generic concepts for blog creation apply to social media.

AI-driven content creators like www.writesonic.com and www.Copy.AI help generate engaging post ideas.

Steps to Do It (Facebook):

1. **Enter Your Content Theme:**
   - Define the type of content you want to post:
     - New listings
     - Market updates
     - Homebuyer tips
     - Client testimonials
2. **Customize Post Format and Tone:**
   - AI tools allow customization of tone (e.g., professional, friendly, fun) and structure (short captions, longer educational posts).
   - This platform has a broader user base, encompassing friends, families, and communities.
   - Content is more casual and personal, making it suitable for showcasing property listings, client testimonials, and community events.
3. **Generate and Review Post:**
   - AI drafts a caption with suggested hashtags and emojis for increased engagement.
   - Visual content like photos, videos, and live streams garner more engagement on Facebook.
   - AI tools that aid in creating eye-catching visuals, editing videos, and generating engaging captions can enhance content effectiveness.
4. **Enhance with AI-Generated Images and Videos:**
   - Tools like www.canva.com offer AI-powered image editing.
   - AI-driven video platforms like www.capcut.com help create short-form content for Facebook Reels.
   - Tools like ChatGPT can assist in summarizing articles for social media posts, extracting key bullet points, and crafting engaging language with appropriate emojis.
5. **Schedule and Monitor Engagement**
   - AI scheduling tools like www.publer.com optimize post timing based on audience activity.
   - AI can analyze post-performance, helping refine future content strategies.
   - Here's a video on the top five social media analytics: https://youtu.be/Cw22bCgrdLU?si=Tm-EtuFO1PhQRRBe

Benefits for Real Estate Agents:

- AI can quickly produce drafts, allowing you to focus on other tasks.
- While a professional tone is important, there's more flexibility for creativity and humor on Facebook.
- AI tools that generate engaging and entertaining content can be utilized more freely.

Considerations:

- **Human Touch is Key**: AI-generated content should be personalized to maintain authenticity.
- **Fact-Checking Required**: AI suggestions may not always be accurate; always review before posting.
- **Avoid Overuse**: Clients appreciate originality; mix AI-generated content with personal updates and real-life interactions.

Social Media Content Generation – LinkedIn

Status:	**Now**
Your priority:	**Medium to High**
Hype v. Reality:	**Reality**
Effort:	**Easy** (for just content creation)
Examples:	Any AI tool that you like, such as Gemini, ChatGPT, Claude www.write.homes AI for real estate marketing www.RealEstateContent.ai

Creating engaging content for LinkedIn can significantly enhance your professional presence and connect you with potential clients in the real estate industry. AI tools can assist in generating compelling LinkedIn posts tailored to your audience.

- If you are currently using LinkedIn, this will accelerate your efforts.
- If you are not but have wanted to, this might be what you need to get started in a bigger way.
- If you are not using LinkedIn (much) and it's not really a part of your plan, you can safely skip this idea.

Steps to Do It (LinkedIn):

1. **Choose an AI Tool:** Select an AI content generator suitable for LinkedIn posts. All the platforms above would be a good choice.
2. **Input Your Preferences:**

   - Provide the tool with details about your target audience, desired tone (e.g., professional, friendly), and any specific topics or keywords you want to include.
     - Primarily a professional networking site, LinkedIn's audience consists of industry professionals, business leaders, and job seekers.
     - Content here should be tailored to a professional tone, focusing on industry insights, market trends, and thought leadership.
3. **Generate Content:**
   - Use the AI tool to create a draft of your LinkedIn post.
   - Review the generated content to ensure it aligns with your objectives and make any necessary adjustments.
   - Emojis and a lighter tone might be better on Facebook, with a slightly more professional demeanor on Linked In.
   - Long-form articles, professional updates, and industry news perform well.
   - AI tools that assist in creating detailed market analyses, informative blog posts, and professional presentations are beneficial.
4. **Schedule or Post:** Once satisfied, you can directly post the content on LinkedIn or schedule it for later using the platform's scheduling features.

Considerations:

- **Authenticity**: Ensure the content reflects your personal voice to maintain authenticity.
- **Accuracy**: Always review AI-generated content to confirm factual correctness and appropriateness.
  - Maintaining a high level of professionalism is crucial.
  - AI-generated content should be thoroughly reviewed to ensure it aligns with professional standards and avoids any informal language or errors.
- **Engagement**: While AI can assist in content creation, personal interaction remains crucial. Engage with your audience by responding to comments and messages to build genuine relationships.

Social Media Content Generation – Instagram

Status: **Emerging**
Your priority: **Low to Medium**
Hype v. Reality: **Midpoint. Not as easy as this write-up is going to suggest.**
Effort: **Medium** (for just content creation)

Example: www.piktochart.com
www.write.homes AI for real estate marketing
www.RealEstateContent.ai

Instagram is a visual-centric platform where eye-catching images and videos are key to engaging your audience. AI tools can assist in creating compelling content tailored for Instagram's unique environment.

How Does It Work?

AI tools for Instagram analyze trends, user preferences, and successful content patterns to help you craft posts that resonate with your target audience. They can suggest design elements, generate captions, recommend hashtags, and even edit photos and videos to enhance visual appeal.

Steps to Do It (Instagram):

1. **Choose an AI Tool**: Select an AI-powered platform designed for Instagram content creation. For instance, www.piktochart.com AI offers an Instagram post generator that crafts captivating visuals from any prompt in seconds. Some of the other websites mentioned above can also be a resource.
2. **Input Your Ideas:** Provide the tool with your concepts, themes, or messages you wish to convey. The AI will use this input to generate tailored content suggestions.
3. **Customize the Content:** Review the AI-generated suggestions and adjust ensure they align with your brand's voice and style. Tools like Canva offer AI-driven templates that you can personalize to maintain brand consistency.
4. **Schedule and Post:** Once satisfied, schedule your posts using the AI tool's scheduling features or post them directly to your Instagram account.

Benefits for Real Estate Agents:

- **Enhanced Visual Appeal**: AI tools can help create high-quality images and videos that capture attention in Instagram's visually driven feed.
- **Time Efficiency:** Automating content creation allows you to maintain a consistent posting schedule without investing extensive time in design and planning.
- **Trend Adaptation**: AI can analyze current Instagram trends and suggest content that aligns with what's popular, increasing the likelihood of engagement.

Considerations:

- **Platform Specificity:**
  - Instagram's focus on high-quality visuals means that content should be aesthetically pleasing and aligned with current design trends.
  - Unlike LinkedIn, which is more text-focused and professional, or Facebook, which balances text and visuals, Instagram demands a strong visual strategy.
- **Audience Engagement**:
  - Instagram users often engage with content through likes, comments, and shares.
  - Ensure your AI-generated content encourages interaction and is tailored to your audience's interests.
- **Brand Consistency:** While AI tools provide suggestions, it's essential to maintain your brand's unique style and voice. Customize AI-generated content to ensure it reflects your brand identity.

By leveraging AI tools tailored for Instagram, you can create visually appealing content that captures your audience's attention and fosters engagement, all while saving time and effort in the content creation process.

Social Media Content Generation – Other Platforms

Facebook Short Videos / TikTok – Assuming it is still available in the U.S.

Status:	**Emerging**
Your priority:	**Low to Medium**
Hype v. Reality:	**Midpoint**
Effort:	**Easy** (for just content creation)
	Medium to Hard (video editing, captions, music selections)

Author Note: We think TikTok is a great platform to reach first time buyers. AI's role is going to be limited in the near term to idea generation and video editing. We're not sure if the idea generation capabilities are really going to help all that much on a platform where creativity is so essential. One way to get started might be to take our 25 Buyer and 25 Seller videos (in five-minute format) and see if you can edit them down to smaller chunks that fit the TikTok model.

- **Overview**: A rapidly growing platform known for short, engaging videos.
- **Audience**: Predominantly younger users, but its user base is diversifying.
- **Content Ideas**: Quick property tours, real estate tips, behind-the-scenes looks, and neighborhood highlights.

- **AI Tools**: AI can assist in editing videos, adding captions, and suggesting trending music to enhance content appeal.

YouTube

Status:	**Now**
Your priority:	**Medium**
Hype v. Reality:	**Reality**
Effort:	**Easy** (for just content creation)
	Medium to Hard (video editing, captions, music selections)

Author Note: We included 50 sample video scripts with links to finished videos for you in this book. If you have ever thought of creating a strong digital presence with video, you now have the tools to do it. Idea generation and content creation is simple. Building the finished videos involves several relatively simple steps, but you'll need to set up and learn all the tools.

- **Overview**: A leading platform for longer-form video content.
- **Audience**: Diverse age range seeking in-depth information.
- **Content Ideas**: Detailed property walkthroughs, client testimonials, market analysis, and educational content about the buying or selling process.
- **AI Tools**: AI can help with video editing, generating transcripts, and optimizing video titles and descriptions for better search visibility.

Pinterest

Status:	**Now** (content)
	Emerging (images, trending topics analysis)
Your priority:	**Low**
Hype v. Reality:	**Low** (content)
	Midpoint (trending topic analysis)
Effort:	**Easy** (for just content creation)
	Medium to Hard (images, trending topics analysis)

Author Note: The visual aspects of real estate play well to this platform. The comments for TikTok apply here too. We suspect that TikTok and Facebook shorts should be a higher priority for agents that use this marketing channel.

- **Overview**: A visual discovery platform where users find inspiration.
- **Audience**: Users interested in home design, decor, and renovation ideas.

- **Content Ideas**: High-quality images of properties, design ideas, and infographics about the real estate process.
- **AI Tools**: AI can assist in creating visually appealing graphics and analyzing trending topics to inform content strategy.

X (formerly Twitter)

Status:	**Now**
Your priority:	**Low**
Hype v. Reality:	**Reality**
Effort:	**Easy** (for just content creation)
	Medium to Hard (scheduling, engagement tracking)

- **Overview**: A platform for real-time updates and engagement.
- **Audience**: Users seeking quick information and industry news.
- **Content Ideas**: Market updates, open house announcements, and sharing relevant articles.
- **AI Tools**: AI can help draft concise posts, schedule tweets for optimal times, and monitor engagement metrics.

Facebook Stories and Reels

Status is like the tools mentioned above.

- **Overview**: Features within Facebook that focus on short-lived or short-form video content.
- **Audience**: Users looking for quick, engaging content.
- **Content Ideas**: Short property highlights, client shout-outs, and quick tips.
- **AI Tools**: AI can aid in creating engaging short videos, adding effects, and analyzing which types of content perform best.

Each platform offers unique opportunities to connect with different audiences. Leveraging AI tools can enhance content creation and engagement across these channels.

Video Content Creation

Status:	**Immediate Reality**
Your priority:	**High**
Hype v. reality:	**Reality**

Effort:	**Easy** (for just content creation) **Medium** (editing, captions, music)
Examples:	Any AI tool that you like, such as Gemini, ChatGPT, Claude www.synthesia.io/features/ai-script-generator https://invideo.io/tools/script-generator https://www.veed.io/ (Create videos) https://www.capcut.com/ (Create videos) https://pictory.ai/ (Create videos with background) https://fliki.ai/ (Create videos with background)

Creating engaging faceless videos for YouTube is now more accessible than ever, thanks to advancements in AI. These videos are particularly useful for real estate agents to provide valuable content to potential buyers without appearing on camera. By leveraging AI tools, you can produce professional-quality videos efficiently. Here's a step-by-step guide to creating such videos.

SCRIPTWRITING

Status:	**Immediate Reality**
Your priority:	**Medium**
Hype v. reality:	**Reality**
Effort:	**Easy** (for just content creation)
Examples:	Synthesia's AI Script Generator InVideo's AI Script Generator

How It Works

- Begin by picking a topic or series of topics. AI can give suggestions on this. Then draft a script that outlines the content of your video. AI-powered script generators can assist in creating engaging narratives tailored to your target audience.

Steps to Do It:

1. **Choose a Script Generator:** Utilize tools like Synthesia's AI Script Generator or InVideo's AI Script Generator.
2. **Input Your Topic:** Enter the main subject or theme of your video.
3. **Generate the Script:** AI will produce a script based on your input, which you can review and edit as needed. You can also look at our Buyer and Seller scripts in this book for ideas (you are free to copy / modify).

Benefits:

- Saves time in content creation.
- Ensures the script is structured and engaging.

Considerations:

- Review the AI-generated script to ensure it aligns with your message and tone.
- Make necessary adjustments to personalize the content and customize it to your market.
- Check to make sure everything is accurate.

VOICE GENERATION

How It Works

- Once your script is ready, you can generate a voiceover without recording your own voice. AI voice generators can produce natural-sounding speech from text.
- Alternatively use a site like Eleven Labs to make copy of your own voice.

Steps to Do It:

1. **Select a Voice Generator:** Use platforms like ElevenLabs or VEED's Text-to-Speech.
2. **Input the Script:** Paste your script into the tool.
3. **Choose a Voice:** Select from various AI-generated voices or create a custom voice clone.
4. **Generate the Voiceover:** The tool will convert your text into speech, which you can download for use in your video.

Benefits:

- Eliminates the need for recording and editing equipment.
- No need to take lots of takes (but there isn't a funny blooper reel either).
- Offers a variety of voice options to match your brand's tone.

Considerations:

- Ensure the pronunciation of specific terms is accurate.
- Some tools may require a subscription for advanced features.

VISUAL CONTENT CREATION

Status:	**Immediate Reality**
Your priority:	**Low**
Hype v. Reality:	**Reality**
Effort:	**Easy** (for just the basics) **Medium** (adding text overlays, transitions and other embellishments)
Examples:	Pictory

Fliki
CapCut

How It Works

- To create a faceless video, you'll need visuals that complement your script.
- AI tools can help generate or source relevant images and clips.

Steps to Do It:

1. **Use AI Video Generators:** Platforms like Pictory and Fliki can transform your script into a video by pairing it with relevant visuals. CapCut is also very popular.
2. **Input Your Script:** Upload your script to the platform.
3. **Select Visuals:** AI will suggest images and clips that match your script's content.
4. **Customize the Video:** Adjust the visuals, add text overlays, and incorporate transitions.

Benefits:

- Streamlines the video creation process.
- Provides access to a vast library of visuals.

Considerations:

- Review the selected visuals to ensure they accurately represent your message.
- Customize elements to maintain brand consistency.

EDITING AND FINALIZING

Status:	**Immediate Reality**
Your priority:	**Low**
Hype v. Reality:	**Reality**
Effort:	**Easy to Medium**
Examples:	VEED InVideo Sora (OpenAI) CapCut

How It Works

- After assembling your visuals and voiceover, finalize your video through editing.

Steps to Do It:

1. **Use Video Editing Tools:** Platforms like VEED and InVideo offer user-friendly interfaces for editing. Tools like Sora (OpenAI) and CapCut can also assist in creating and editing videos efficiently.
2. **Upload Your Content:** Combine your visuals and voiceover.
3. **Add Enhancements:** Incorporate background music, captions, and any additional effects.
4. **Preview and Export:** Review the final product and export it in your desired format.

Benefits:

- Enhances the professionalism of your video.
- Allows for creative customization.

Considerations:

- Ensure all elements are cohesive and align with your brand.
- Keep the video concise to maintain viewer engagement.

AI can translate videos into different languages, helping you reach a broader audience, such as the Hispanic community.

Video Conclusion

By leveraging AI tools, real estate agents can efficiently create engaging faceless videos for YouTube. These videos can effectively showcase properties, provide market insights, or offer valuable tips to potential buyers, all without the need to appear on camera. For a practical demonstration of creating faceless YouTube videos using AI, you might find this video helpful: https://youtu.be/dCfjog4Np4A

AI-Driven Lead Generation and Qualification

Predictive Analytics

Status: **Early Emerging Reality**
Your priority: **Low to Medium**
Hype v. Reality: **Midpoint, but the platform will charge you; you can't do yourself.**
Effort: **Medium**

Examples:	https://offrs.com/	Listing leads
	https://likely.ai/	Identifies likely sellers
	https://revaluate.com/	Identifies likely sellers

AI is transforming how real estate agents find and qualify potential clients. By analyzing large amounts of data, AI can predict which individuals are most likely to buy or sell properties. This process is known as predictive analytics. Irritatingly, you likely won't be able to do this work yourself. Instead, you'll need to pay a company to do it for you. That makes it easier, but it's an on-going expense.

One of our top agents is using https://revaluate.com/. They report they have not gotten listings yet, but they are optimistic and are continuing to use the service. If you like being on the leading edge, consider it. If not, wait a while and see how much it improves.

How It Works

- AI systems examine various types of information, such as:
    - **Property Details:** Information about homes, like their size, age, and location.
    - **Demographics:** Data about people, including age, income, and family size.
    - **Behavior Patterns:** Actions people take, such as visiting real estate websites or attending open houses.
    - **Life Events:** Significant changes like job relocations or family growth.
- By studying these factors, AI can identify patterns and forecast who might be ready to make a real estate transaction. For example, someone frequently searching for homes in a specific area might be considering a move.

Steps to Do It:

1. **Choose an AI Tool:** Select a platform designed for real estate predictive analytics. Some options include:
    - **Offrs:** Uses AI to predict which homeowners are likely to sell soon.
    - **Likely.AI:** Analyzes data to forecast which properties may be listed in the next few months.
2. **Input Data:** Provide the tool with relevant information, such as property records and demographic details.
3. **Analyze Results:** Review the tool's predictions to see which leads are most promising.
4. **Reach Out:** Contact the individuals identified as likely buyers or sellers to offer your services

Benefits:

- **Efficient Use of Time:** Focuses your efforts on the most promising leads, saving time.
- **Improved Marketing:** Allows for targeted campaigns aimed at those most likely to act.
- **Competitive Advantage:** Gives you insights that others without AI tools might miss.

Considerations:

- **Data Privacy:** Ensure that all data used complies with privacy laws and respects individual confidentiality.
- **Accuracy:** While AI provides valuable predictions, it's essential to combine these insights with your personal expertise and judgment.
- **Cost:** Some AI tools can be expensive, so consider your budget when selecting a platform.

By incorporating AI-driven predictive analytics into your real estate practice, you can more effectively identify and engage with potential clients, enhancing your chances of success in a competitive market.

Chatbots and Virtual Assistants

Status:	**Early Emerging Reality**
Your priority:	**Very Low**
Hype v. Reality:	**Closer to Hype**
Effort:	**Hard**
Examples:	https://landbot.io/

AI chatbots are like virtual helpers that can talk to people on your website or through messaging apps. In real estate, these chatbots can answer questions, gather information from potential clients, and help schedule appointments, all without needing a person to be available 24/7. This means you can assist visitors at any time, even when you are busy, or it is late at night. We think this is going to be for the agent that is doing a lot of marketing to drive traffic to their website. The bot can help with conversion if a live agent is not available. It's a backup plan. Most real estate agents likely don't need this.

How It Works

When someone visits your real estate website, the chatbot pops up to greet them. It can ask questions like, "Are you looking to buy or rent?" or "What area are you interested in?" Based on their answers, the chatbot provides information about properties that match their needs. It can also collect their contact details for you to follow up later. Some advanced chatbots can even schedule property viewings by checking your calendar.

Steps to Set It Up:

1. **Choose a Chatbot Platform:** There are several platforms designed for real estate agents. For example, Landbot allows you to create custom chatbots without any coding in under 30 minutes.
2. **Customize the Chatbot:** Use the platform's tools to set up the chatbot's conversation flow. Decide what questions it should ask and how it should respond based on different answers. For instance, if a visitor is interested in buying, the chatbot can ask about their preferred location and budget.
3. **Integrate with Your Website or Messaging App:** Once your chatbot is ready, add it to your website or connect it to a messaging app like Facebook Messenger. This usually involves copying and pasting a code snippet provided by the chatbot platform.
4. **Test the Chatbot:** Before going live, test the chatbot to ensure it works as expected. Pretend to be a visitor and go through different conversation paths to see if everything functions correctly.
5. **Monitor and Improve:** After launching, keep an eye on how the chatbot performs. Many platforms offer analytics to show you how users interact with the chatbot. Use this information to make improvements over time.

Benefits:

- **Always Available:** The chatbot can assist visitors at any time, ensuring no potential lead is missed, even outside of regular business hours.
- **Quick Responses:** Visitors get immediate answers to their questions, enhancing their experience on your site.
- **Lead Qualification:** By asking key questions, the chatbot can determine how serious a visitor is and gather their contact information for follow-up.
- **Timesaving:** Automating initial interactions frees up your time to focus on more critical tasks, like closing deals.

Considerations:

- **Keep It Simple:** Ensure the chatbot's language is easy to understand. Avoid complex terms that might confuse visitors.
- **Regular Updates:** As your property listings change, update the chatbot to provide accurate information.
- **Human Backup:** While chatbots are helpful, some visitors may prefer speaking to a person. Ensure there's an easy way for the chatbot to hand over the conversation to you or a team member when needed.

- **Privacy Matters:** Be transparent about data collection. Inform visitors that their information will be used to assist them and will be kept confidential.

By incorporating an AI chatbot into your real estate business, you can enhance customer engagement, capture more leads, and streamline your operations, all while providing a seamless experience for your website visitors.

AI-Generated Real Estate Marketing Emails

Status:	**Early Emerging Reality**
Your priority:	**Very Low**
Hype v. Reality:	**Closer to Hype**
Effort:	**Hard**
Examples:	ChatGPT (openai.com) Jasper AI (jasper.ai) HubSpot AI Email Tool (hubspot.com ActiveCampaign (activecampaign.com) Mailchimp (mailchimp.com) Brevo (brevo.com) Seventh Sense AI (seventhsense.ai)

How It Works:

Email marketing is one of the most effective ways for real estate agents to stay in touch with leads, past clients, and prospects. AI-powered tools can write, personalize, and automate emails, saving time while improving engagement. Whether you're following up with a new lead, nurturing a long-term client, or sending a market update, AI ensures your emails are well-written, relevant, and timed for maximum impact.

Steps to Set It Up:

1. **Choose an AI Email Generator:**
   - **ChatGPT** (openai.com) – Great for drafting real estate emails with customizable tone and style.
   - **Jasper AI** (jasper.ai) – Writes compelling emails tailored for marketing and sales.
   - **HubSpot AI Email Tool** (hubspot.com) – Automates email campaigns and personalizes messaging.
2. **Define Your Audience and Goal:**
   - **New leads** – AI can craft an engaging introduction email explaining your services.

- **Past clients** – Send **a** personalized follow-up about recent home values in their area.
- **Newsletter subscribers** – Share monthly market updates or neighborhood spotlights.
- Example: If a lead inquires about homes in Boulder, CO, AI can generate a personalized email with recent listings, pricing trends, and next steps.

3. **Generate and Customize the Email Content:**
    - AI suggests subject lines to boost open rates.
    - Example: Instead of "Check Out These Listings," AI might generate "Just Listed: Stunning Homes in Boulder You'll Love!"
    - AI writes persuasive content based on your goal (introducing yourself, showcasing listings, offering real estate tips, etc.).
4. **Personalize with AI and CRM Integration:**
    - AI tools like ActiveCampaign (activecampaign.com) and Mailchimp (mailchimp.com) pull client names, saved searches, and past interactions to personalize emails.
    - Example: AI can auto-fill a message with a client's name, preferred price range, and recent search history.
5. **Automate the Email Campaign:**
    - AI helps schedule follow-ups at the right time to increase response rates.
    - Example: If a lead opens your email but doesn't reply, AI can send a reminder a few days later with a "Just Checking In" message.
    - Tools like **Brevo** (brevo.com) allow real estate agents to set up automated drip campaigns (e.g., a 5-email sequence to nurture new leads).
6. **Analyze and Optimize:**
    - AI tracks open rates, clicks, and responses to see which emails perform best.
    - Tools like Seventh Sense AI (seventhsense.ai) suggest the best time to send emails for each contact based on their past engagement.

Benefits:

- Saves time by writing and automating emails.
- Increases open rates and engagement with personalized messages.
- Helps real estate agents stay top-of-mind with potential buyers and sellers.
- AI-driven follow-ups increase conversion rates.

Considerations:

- AI-generated emails should still feel human—always review before sending.

- Too much automation can feel impersonal—mix AI with real conversations.
- Ensure emails comply with spam laws (CAN-SPAM Act, GDPR, etc.).

With AI-powered email marketing, real estate agents can attract, nurture, and convert leads more efficiently, while maintaining a personal touch.

AI for Lead Scoring and Prioritization

Status:	**Early Emerging Reality**
Your priority:	**Very Low**
Hype v. Reality:	**Closer to Hype**
Effort:	**Hard**
Examples:	Follow Up Boss (followupboss.com) LionDesk (liondesk.com) Revaluate (revaluate.com) SmartZip (smartzip.com) ActiveCampaign Brevo

How It Works:

Not all leads are equal. Some are ready to buy or sell now, while others may take months or even years. AI-powered lead scoring helps real estate agents identify the most promising leads by analyzing engagement, behavior, and market data. AI can track website visits, email opens, social media interactions, and even predictive factors like life changes and financial readiness to rank leads from "hot" to "cold." This allows real estate agents to focus on high-value prospects while nurturing long-term opportunities.

Steps to Set It Up:

1. **Use an AI-Powered CRM for Lead Tracking:**
   - Platforms like Follow Up Boss (followupboss.com) and LionDesk (liondesk.com) track client activity, such as email engagement, property searches, and call history.
   - AI assigns each lead a **score** based on their level of interest and interaction.
2. **Set Up AI-Based Lead Scoring Rules:**
   - AI assigns higher scores to leads who:
     - Frequently open emails or click property links.
     - Request more details or schedule a showing.
     - Engage on social media (likes, shares, comments).
     - Use home valuation tools (indicating seller interest).

- Leads who don't engage may be assigned lower scores and placed into long-term nurture campaigns.

3. **Identify Predictive Selling and Buying Signals:**
   - AI-powered tools like Revaluate (revaluate.com) analyze public records, job changes, and life events (e.g., marriage, new baby, downsizing) to predict who is most likely to move soon.
   - SmartZip (smartzip.com) uses AI to analyze homeowner data and local trends to predict which properties may hit the market next.
4. **Prioritize High-Intent Leads for Immediate Follow-Up:**
   - AI highlights leads that show strong buying or selling intent so real estate agents can respond quickly before they lose interest.
   - Example: A lead who views the same listing five times in a week might trigger an AI-generated alert:
     - "John Doe has viewed [Property Address] five times. Consider reaching out with a personalized message or a tour offer."
5. **Automate Lead Nurturing for Low-Priority Leads:**
   - AI-powered email tools like ActiveCampaign and Brevo automate check-ins with leads who aren't ready to act.
   - Example: AI can place "cold" leads into a 6-month drip campaign with useful content:
     - Month 1: "Understanding Interest Rates for Homebuyers"
     - Month 3: "What's Happening in Your Local Market?"
     - Month 6: "Is It Time to Start Your Home Search?"
6. **Track and Adjust Lead Scoring Over Time:**
   - AI learns from past transactions to refine scoring models over time.
   - If leads with a certain behavior (e.g., using mortgage calculators) frequently convert into clients, AI can assign higher scores to similar future leads.

Benefits:

- Saves time by prioritizing high-value leads.
- Increases conversion rates by following up with engaged prospects first.
- Helps predict future sellers and buyers, giving agents a competitive edge.
- Keeps long-term leads engaged without manual effort.

Considerations:

- AI predicts behavior but isn't perfect—agents should still use their judgment.
- Lead scoring is only as good as the data—ensure CRM tools are updated regularly.
- Cold leads shouldn't be ignored—some take longer to move from interest to action.

AI-powered lead scoring helps real estate agents work smarter, not harder, by focusing on leads who are most likely to convert.

Lead Generation and Qualification Conclusion

By leveraging AI in lead generation, qualification, marketing, and prioritization, real estate agents can streamline their efforts and maximize success. Some AI tools have the capability to predict seller and buyer behavior, automate follow-ups, and enhance engagement, providing the agent a competitive edge in an increasingly digital market. AI tools like chatbots and email automation platforms might further streamline client interactions, allowing agents to respond quickly and efficiently, even outside regular business hours. However, while AI provides valuable insights, it's important for agents to combine these tools with their expertise and maintain a personal touch. AI should be used as a complement to, not a replacement for, relationship-building. As technology continues to evolve, agents who strategically integrate AI into their workflow will be better positioned to stay ahead in the industry.

Enhancing Client Communication and Relationship Management

Personalized Outreach

Status:	**Emerging Reality**
Your priority:	**Low**
Hype v. Reality:	**Hype**
Effort:	**Very Hard**
Examples:	https://www.glideapps.com/agents/real-estate

Author Note: We included this, since anytime someone asks what AI can do for real estate agents, something like this comes up. Personally, we think this seems like a much lower value use of AI than some of the prior examples. There likely are a small number of specialist real estate agents with a unique business plan that can really use this – this is for you!

Author Note: Most of you can skip this one.

AI is transforming how real estate agents connect with clients by enabling personalized communication. This means tailoring messages to fit each client's unique needs and interests, making them feel valued and understood.

How It Works:

AI tools analyze data about your clients, such as their property preferences, search history, and past interactions. By understanding this information, AI can help you craft messages that speak directly to each client's interests. For example, if a client has been looking at homes with large kitchens, AI can suggest sending them listings that highlight spacious cooking areas.

Steps to Set It Up:

1. **Choose an AI Tool:** Select a platform designed for personalized outreach in real estate. For instance, Glide offers AI agents that handle tasks like lead generation and follow-up emails.
2. **Gather Client Data:** Collect information about your clients' preferences, such as the types of properties they are interested in, preferred locations, and any specific features they desire.
3. **Set Up Personalized Messages:** Use the AI tool to create messages tailored to each client's interests. For example, if a client is interested in homes near good schools, the AI can help draft an email showcasing properties in such areas.
4. **Automate Outreach:** Schedule these personalized messages to be sent at optimal times. Some AI tools can determine the best times to reach each client based on their past interactions.
5. **Monitor and Adjust:** After sending out messages, track how clients respond. If certain types of messages receive more positive feedback, adjust your strategy accordingly.

Benefits:

- **Enhanced Client Engagement:** Personalized messages show clients that you understand their needs, making them more likely to engage with you.
- **Time Efficiency:** Automating personalized outreach saves time, allowing you to focus on other important tasks.
- **Higher Conversion Rates:** Clients are more likely to proceed with a purchase when they feel the communication is tailored to them.

Considerations:

- **Data Privacy:** Ensure that you handle client data responsibly and comply with privacy laws.
- **Authenticity:** While AI can assist in crafting messages, it's important to maintain a genuine tone to build trust with clients.
- **Continuous Learning:** Regularly update the AI tool with new client data to keep the outreach relevant and effective. By leveraging AI for personalized outreach, real estate agents can build stronger relationships

with clients, leading to increased satisfaction and success in the competitive housing market.

Customer Relationship Management (CRM)

Status:	**Now**
Your priority:	**Low to High**
Hype v. Reality:	**Midpoint**
Effort:	**Medium**
Example:	kvCORE

If you get most of your transactions networking or repeat and referral, you likely don't need an AI-enabled CRM. (**Author Note:** Most of you should stick with what you are doing now.) There is a lot of hype in this market segment. If you get most of your transactions from lead generation, especially online lead generation, then this section is for you

AI is transforming how real estate agents manage their relationships with clients through advanced Customer Relationship Management (CRM) systems. These AI-powered CRMs help agents organize client information, track interactions, and automate tasks, leading to more efficient and personalized service.

How It Works:

AI-powered CRMs collect and analyze data from various sources, such as emails, phone calls, and social media interactions. The AI processes this information to provide insights into client behavior and preferences. For example, if a client frequently searches for homes in a particular neighborhood, the CRM can highlight this interest, allowing the agent to tailor their communications accordingly.

Steps to Implement AI-Powered CRM:

1. **Select a Suitable CRM Platform:** Choose a CRM that integrates AI features tailored for real estate. Platforms like kvCORE offer AI-driven tools designed to convert leads and engage clients effectively.
2. **Input Client Data:** Gather and enter comprehensive information about your clients into the CRM, including contact details, property preferences, and communication history.
3. **Utilize AI Features:** Leverage the AI capabilities of the CRM to analyze client data. For instance, the system can identify patterns in client behavior, such as preferred property types or communication channels. You hear this a lot, but it's something that will be at the big platform level. Don't expect your usage of AI is going to predict client behavior!
4. **Automate Tasks:** Set up the CRM to automate routine tasks like sending follow-up emails, scheduling appointments, or reminding you of important dates, ensuring consistent and timely communication with clients.

5. **Monitor and Adjust:** Regularly review the insights and suggestions provided by AI to refine your client engagement strategies. Adjust your approach based on the CRM's recommendations to enhance client satisfaction.

Benefits:

- **Personalized Client Interactions:** AI analyzes client data to help you tailor your communications, making clients feel valued and understood.
- **Increased Efficiency:** Automating routine tasks frees up time, allowing you to focus on building relationships and closing deals.
- **Improved Lead Management:** AI can prioritize leads based on their likelihood to convert, enabling you to focus your efforts where they matter most.

Considerations:

- **Data Privacy:** Ensure that all client data is handled in compliance with privacy laws and regulations.
- **System Integration:** Verify that the CRM integrates seamlessly with your existing tools and workflows to avoid disruptions.
- **Training and Support:** Choose a CRM provider that offers adequate training and support to help you maximize the system's capabilities. By adopting an AI-powered CRM, real estate agents can enhance their client communication and relationship management, leading to increased client satisfaction and business growth.

AI-Powered Social Media Scheduling

Status:	**Now**
Your priority:	**Low**
Hype v. Reality:	**Reality**
Effort:	**Medium to Hard**

How It Works:

Posting consistently on social media is critical for real estate agents, but it can be time-consuming. AI-powered scheduling tools help by planning, optimizing, and automating posts across platforms like Facebook, Instagram, LinkedIn, and Twitter (X). These tools can suggest content ideas, generate captions, schedule posts at the best times, and even analyze engagement to improve future content.

Author Note: For the small number of real estate agents with extensive social media campaigns, this should be on your to-do list. Otherwise, this is overkill for most.

Steps to Do It:

1. **Choose an AI-Powered Scheduling Tool:**
   - **Buffer** (buffer.com) – Schedules posts across multiple platforms and suggests the best posting times.
   - **Hootsuite** (hootsuite.com) – Automates social media planning and provides analytics on post performance.
   - **Later** (later.com) – Specializes in **Instagram-first** scheduling, helping real estate agents optimize image-based posts.
   - **Metricool** (metricool.com) – Offers AI-generated post suggestions, hashtag research, and post-performance tracking.
2. **Generate Content with AI:**
   - Use AI tools like ChatGPT, Jasper AI, or Canva Magic Write to create captions, hashtags, and post descriptions.
   - Example: AI can turn a market update report into a clear, engaging post.
     - Raw data: "Median home price in Denver increased by 2.1% year-over-year to $638,000."
     - AI-generated post: "Denver's real estate market is holding strong! Median home prices are up 2.1% from last year, now at $638,000. More homes are available, making this a great time to explore your options."
3. **Schedule Posts in Advance:**
   - Upload posts to the scheduling platform and let AI suggest the best time to post based on engagement trends.
   - Most platforms allow bulk scheduling, so agents can plan a month of content in one sitting.
4. **Monitor and Adjust with AI Analytics:**
   - AI analyzes which posts perform best and suggests ways to improve engagement.
   - Example: If AI detects that listing photos with pricing details perform better than general real estate tips, it will suggest shifting content in that direction.
5. **Engage with Followers Using AI:**
   - Some tools offer AI-powered auto-responses to quickly reply to common comments or messages.
   - Example: A chatbot can automatically respond to inquiries about a property and route serious buyers to your inbox.

Benefits:

- Saves hours per week by automating posting and scheduling.
- Ensures consistent content, even on busy days.
- Uses data-driven insights to refine your strategy and improve engagement.

- Helps you maintain a presence across multiple platforms without extra effort.

Considerations:

- AI can help, but engagement still requires a personal touch—respond to comments yourself when possible.
- Different platforms require different content styles (e.g., Instagram is more visual, LinkedIn is more professional).
- Automated scheduling should not make your posts feel robotic—ensure variety in tone and content.

AI-powered social media scheduling allows real estate agents to stay visible, attract leads, and build brand awareness without spending hours online.

Referral and Past Client Marketing

Status:	**Emerging Reality**
Your priority:	**Low to Medium today; higher soon**
Hype v. Reality:	**Closer to Hype now, but likely to change soon**
Effort:	**Hard**
Examples:	Follow Up Boss (followupboss.com) Wise Agent (wiseagent.com) Homebot (homebot.ai) ActiveCampaign Brevo Hootsuite Buffer

How It Works:

Most real estate business comes from repeat clients and referrals, but staying in touch consistently can be challenging. AI can help by automating personalized outreach, tracking client milestones, and reminding you when to follow up. AI-powered tools can also analyze client interactions and suggest the best ways to re-engage past buyers and sellers.

Author Note: You might have to change to a new CRM to enable this strategy. It's not clear to the authors that it is worth it to do it now. It is likely your current CRM will also add these features soon. If you see a really clear, big benefit, jump to a new system now; else wait.

Steps to Do It:

1. **Use AI to Track Past Clients and Milestones:**

- AI-powered Customer Relationship Management (CRM) tools like Follow Up Boss (followupboss.com) and Wise Agent (wiseagent.com) track client anniversaries, birthdays, and home purchase dates.
- Example: AI can remind you when a past client's home purchase anniversary is approaching so you can send a congratulatory email or card.

2. **Automate Personalized Check-Ins and Updates:**
   - AI tools like ChatGPT and Jasper AI can generate warm, personal messages that sound natural, not robotic.
   - Example: If a client bought a home in 2021, AI can generate an email like:
     - "Can you believe it's been three years since you bought your home? Hope you love it! Let me know if you need any real estate advice or just want to catch up."
   - AI can also schedule follow-ups every 6-12 months, so past clients always hear from you first when thinking about moving again.
3. **Send AI-Generated Market Updates to Past Clients:**
   - Use tools like Homebot (homebot.ai) to send personalized home value updates automatically.
   - Example: AI can generate a monthly market update showing how much equity a client has gained since their home purchase.
4. **Automate Referral Requests with AI:**
   - AI can track which past clients engaged positively (e.g., left a review, responded to emails) and suggest the best time to ask for a referral.
   - Example: A CRM might flag a past client who left a 5-star review, prompting you to send:
     - "It was a pleasure helping you find your home! If you know anyone looking to buy or sell, I'd love an introduction."
   - AI-powered email tools like ActiveCampaign or Brevo can set up automated referral request sequences that don't feel pushy.
5. **Engage on Social Media with AI Assistance:**
   - AI-powered scheduling tools like Hootsuite and Buffer can help maintain relationships by sharing relevant content for past clients.
   - Example: AI can suggest seasonal home maintenance tips, such as:
     - "Winter is here! Have you checked your furnace filter? Here are some easy ways to prep your home for colder weather."

Benefits:

- Ensures consistent communication with past clients.
- Increases repeat business and referrals with minimal effort.
- Saves time by automating milestone reminders and outreach.
- Provides valuable, personalized content instead of generic messages.

Considerations:

- AI helps automate outreach, but a personal touch still matters—add custom notes when possible.
- Be mindful of spam laws when automating email sequences.
- Not all past clients want frequent communication—AI can track engagement and suggest who to contact most.

Author Note: Again, this sounds great, almost like science fiction, but the authors are skeptical that it is this broadly available.

Client Communication and Relationship Management Conclusion

By using AI for referral and past client marketing, real estate agents can stay top-of-mind effortlessly and generate more business from their existing client base. Agents can use AI to tailor messages based on client preferences, past interactions, and property interests, increasing client engagement and conversion rates. AI-powered CRMs streamline managing client data, track interactions, and automate tasks, allowing agents to provide more efficient and customized service.

SEO Optimization and Market Analysis

Search Engine Optimization (SEO) makes your website easier to find on the internet. It's about using the right words and making sure your site works well so that search engines like Google can find and show it to people.

Artificial Intelligence (AI) can help with SEO by doing things like:

- Finding popular words people search for.
- Helping write content that includes these words.
- Looking at how people use your site and suggesting improvements.

By using AI, businesses can make their websites more visible, attract more visitors, and provide better information to users.

Search Engine Optimization (SEO)

Status:	**Now**
Your priority:	**Low** (for the majority of agents)
Hype v. Reality:	**Reality**
Effort:	**Hard**

How It Works:

AI-powered SEO tools analyze what people are searching for and help optimize your content by suggesting keywords, meta descriptions, headlines, and blog topics.

For real estate agents, AI can help improve local SEO, ensuring your website appears when people search for "real estate agent near me," "homes for sale in [city]," or "best real estate agent in [neighborhood]."

Author Note: It's a small fraction of agents that will want to dig into this. For the minority, it's a powerful tool.

Steps to Do It:

1. **Find Relevant Keywords:**
   - Use AI-powered keyword research tools like Semrush (semrush.com) or Ahrefs (ahrefs.com) to identify popular search terms in your market.
   - Free tool: Google's "People Also Ask" section and AnswerThePublic (answerthepublic.com) show common real estate-related searches.
2. **Optimize Your Website Content:**
   - AI tools like Surfer SEO (surferseo.com) analyze top-ranking real estate websites and suggest content improvements.
   - Use ChatGPT or Jasper AI to rewrite your property descriptions, bio, and blog posts with better keyword placement.
   - Example: Instead of "Great home in Denver," an AI might suggest, "Spacious 3-bedroom home for sale in Denver near Cherry Creek."
3. **Improve Blog and Social Media SEO:**
   - AI tools like RankIQ (rankiq.com) help create blog topics that are easy to rank on Google.
   - Pictory (pictory.ai) and Canva (canva.com) help turn blogs into engaging social media posts with SEO-friendly captions.
   - Example: If "Best Neighborhoods in Denver" is trending, AI can help you create a detailed blog post and summarize it for social media.
4. **Enhance Local SEO:**
   - Claim and optimize your Google Business Profile (google.com/business) to appear in local searches.
   - AI tools like Yext (yext.com) keep your contact info and listings consistent across directories, improving search rankings.
   - Example: AI can suggest responding to Google Reviews with relevant keywords (e.g., "Thank you for choosing me as your Denver real estate agent!").
5. **Monitor and Improve SEO Performance:**

- Use Google Search Console (search.google.com/search-console) to track your rankings.
- AI-powered tools like Moz Pro (moz.com) suggest ways to improve your content and fix broken links.

Benefits:

- Higher rankings on Google bring more organic leads without paid ads.
- Saves time by automating keyword research and content optimization.
- Improves local search visibility, so more people in your area find your services.

Considerations:

- SEO takes time—results don't happen overnight, but AI speeds up the process.
- AI-generated content should be reviewed for accuracy and a human touch.
- Consistency is key—AI tools help, but regular updates keep you ranking high.

Market Reports

Status:	**Early Emerging**
Your priority:	**Very Low**
Hype v. Reality:	**Hype**
Effort:	**Medium**
Examples:	Altos Research (altos.re)
	HouseCanary (housecanary.com)
	Remine (remine.com)

How It Works:

Understanding and communicating market trends is essential for real estate agents. AI tools analyze vast amounts of data—including MLS listings, sales history, mortgage rates, and economic indicators—to generate insights on pricing trends, supply and demand, and buyer activity. AI can also help turn complex data into clear, compelling reports to share with clients.

Author Note: We've tried it. We have not yet seen MLS data cited. We have seen sites like Redfin and Reddit. We'd trust Redfin to do a good job, but certainly not Reddit! Another danger: If you ask, "What are the real estate market trends in Washington Park, in Denver CO", you won't know if your geographic boundaries for Washington Park are the same as Redfin used. It's also possible that your AI could merge information from multiple sources, all of which have a different geographic boundary and data analysis methodology. We'd suggest caution using AI for market reports and/or analysis.

Steps to Do It:

1. **Use AI-Powered Market Analysis Tools:**
   - **Altos Research** (altos.re) – Provides real-time market trend data and inventory analysis.
   - **HouseCanary** (housecanary.com) – Uses AI to forecast home price appreciation and market conditions.
   - **ChatGPT or Jasper AI** – Helps summarize reports and generate client-friendly explanations.
   - **MLS Data + AI** – Many MLS systems now integrate AI to generate reports on neighborhood trends and pricing shifts.
2. **Gather Market Data for Your Target Area:**
   - AI tools scan recent home sales, price changes, time on market, and mortgage rate trends.
   - Example: Instead of manually analyzing dozens of listings, AI can summarize, "The median home price in Denver rose 2.1% year-over-year, while condos saw slower appreciation."
3. **Generate Easy-to-Understand Market Reports:**
   - AI can convert raw data into digestible insights for buyers and sellers.
   - Example: AI-generated text for a client email.
     - "Home inventory in Denver increased 11.6% last month, giving buyers more options. Prices are stable, but homes are spending longer on the market."
   - AI-powered platforms like Homebot or Altos Research create automated monthly market reports for your sphere of influence.
4. **Customize Market Reports for Specific Clients:**
   - AI tools can generate personalized insights based on a client's home or neighborhood.
   - Example: If a past client bought a home in 2020, AI can show them how much their home value has increased and whether now is a good time to sell.
   - Tools like Remine (remine.com) analyze public records and homeowner data to predict who may be likely to sell soon.

   Author Note: That sounds great—let's predict who will sell next! When we have tried it, we usually get a variation of: "Identifying specific homeowners in Washington Park who may be likely to sell soon is challenging due to privacy considerations and the dynamic nature of the real estate market."
5. **Use AI to Create Social Media & Blog Content:**
   - AI can generate monthly market update posts for Facebook, LinkedIn, and Instagram.

    - Example: AI-generated LinkedIn post. "Denver Real Estate Update: Prices are steady, but inventory is rising. Buyers, this is your chance to negotiate better deals! See the latest trends here."
- Tools like Canva Magic Write and ChatGPT help format market data into engaging visuals and infographics.

Benefits:

- Saves time by automatically analyzing and summarizing market trends.
- Provides real-time insights that help agents advise clients.
- Enhances marketing by turning dry statistics into engaging blog posts.
- Improves lead generation by proactively identifying potential clients.

Considerations:

- AI-generated reports should always be fact-checked against MLS data.
- Some AI insights are predictive, not guaranteed—use them as guidance, not absolutes.
- Not all clients understand market data—AI helps simplify reports, but you should still be available to explain them.

AI-powered market trend analysis helps real estate agents stay ahead of shifts in the market, create valuable content, and provide clients with expert guidance—all with less manual effort.

Author Note: We see breathless, exclamation-mark laden phrases like that all over the internet. We're not buying it.

Optimizing Your Online Presence, Social Proof and Reputation Management

Building trust as a real estate agent starts with a strong online presence. By optimizing your profiles, showcasing social proof, and managing your reputation, you increase visibility and credibility—making it easier to attract and retain clients.

Optimizing Your Online Presence

A well-optimized online presence ensures potential clients can easily find and connect with you. AI-powered tools can help improve your visibility, branding, and engagement across key platforms.

- **Example:** AI-generated Instagram bio: *"Helping first-time home buyers in Denver find their dream home. DM for free homebuyer tips!"*

Social Proof

Social proof and reputation management both help real estate agents build trust, but they work in different ways.

Social proof is about showing that others trust and recommend you. It includes client testimonials, online reviews, success stories, and social media engagement (likes, shares, and comments). When potential buyers or sellers see that others have worked with you and had a great experience, they're more likely to trust you.

- **Example**: A past client posts, "Best real estate agent ever! Helped us find our dream home in just two weeks!" on Google Reviews or Zillow. That positive feedback can influence new clients.

Reputation Management

Reputation management is about protecting and improving your online image. This includes responding to reviews (both good and bad), fixing incorrect information online, and ensuring your website and social media look professional. If a client leaves a negative review, a professional response can help protect your reputation.

- **Example**: A client complains about slow responses. You reply, "I'm sorry for the delay and appreciate your feedback. I'd love to make it right."

Both are essential for attracting and keeping clients—social proof brings in leads, while reputation management ensures you stay credible and trustworthy.

Optimizing Your Online Profile

Status:	**Emerging**
Your priority:	**High**
Hype v. Reality:	**Midpoint**
Effort:	**Easy**
Examples:	SurferSEO BrightLocal RiteTag Flick AI

Author Note: AI-driven online presence optimization can significantly improve visibility and lead generation. While not an immediate priority for all agents, those looking to scale their business should consider implementing these strategies.

How It Works:

AI-powered tools can optimize your online presence by enhancing your Google Business Profile (GBP), improving social media engagement, and ensuring consistent branding. AI can suggest high-ranking keywords, generate engaging content, and streamline review management, making it easier to attract and convert leads.

Steps to Do It:

1. **Optimize Your Google Business Profile with AI**
   - Use AI-powered tools like SurferSEO, BrightLocal, or ChatGPT to suggest high-ranking keywords for your business description and posts.
   - AI-generated business descriptions can highlight your expertise, such as:
     - *"Helping first-time home buyers in Denver navigate the market with confidence. Let's find your dream home!"*
   - Keep your Q&A section fresh by using AI to anticipate common buyer and seller questions.
2. **Enhance Social Media Profiles with AI**
   - **AI-Optimized Bios and Headlines:** AI tools like ChatGPT, RiteTag, and Flick AI can generate compelling, keyword-rich bios.
     - Example: *"Denver's go-to agent for first-time home buyers. DM for a free consultation!"*
   - **AI-Generated Hashtags:** AI analyzes industry trends and suggests high-performing hashtags to increase post visibility.
   - **Professional Cover Photos and Branding**: AI-driven design tools like Canva AI create polished graphics that maintain a consistent brand identity.

Social Proof in Real Estate

Status:	**Emerging**
Your priority:	**Low**
Hype v. Reality:	**Midpoint**
Examples:	Boast.io (boast.io) Trustpilot (trustpilot.com) Brand24 (brand24.com)

Author Note: This is often cited as a powerful new use case for AI. We think for most agents, this will be a lower priority for study and implementation.

How It Works:

AI-powered tools can automate the collection of testimonials, analyze client sentiment, and generate compelling content from reviews and social media interactions. This makes it easier to highlight why potential buyers and sellers should work with you.

Steps to Do It:

1. **Automate the Collection of Client Testimonials:**
   - Use AI-powered tools like Boast.io (boast.io) or Trustpilot (trustpilot.com) to automatically request reviews from past clients via email or text.
   - AI can suggest personalized prompts like:
     - "Hi [Client's Name], congratulations on your home purchase! If you had a great experience, we'd love a short review to help future buyers."
2. **Turn Reviews into Marketing Content:**
   - AI tools like Canva Magic Write and Jasper AI can summarize long client testimonials into engaging social media posts or website highlights.
   - Example:
     - **Review:** "Sarah was so patient in helping us find the perfect home! She knew the market well and negotiated a great deal."
     - **AI-Generated Post:** "Another happy homeowner! 'Sarah's market expertise got us an amazing deal!' Thinking of buying? Let's talk!"
3. **Generate Social Proof Graphics and Videos:**
   - AI-powered design tools like Canva (canva.com) and Animoto (animoto.com) create professional-looking graphics and video testimonials.
   - Example: Turn client success stories into short video reels with before-and-after photos of homes.

   Author Note: While this might not happen with many of your clients, it does seem like a great idea when it does work. Or if you work with fix and flip investors, this would be a great strategy.
4. **Use AI to Monitor and Engage with Social Proof:**
   - AI tools like Brand24 (brand24.com) track mentions of your name or brokerage online, making it easy to highlight positive feedback and respond to any concerns.
   - AI chatbots can automatically thank clients for reviews and suggest responses to frequently asked questions.

   Author Note: We're skeptical this is worth the effort at the agent level, but for a large brokerage, it could be important.
5. **Leverage AI-Generated Social Proof for Website SEO:**
   - AI can analyze which testimonials perform best and suggest which ones to feature on your website for maximum engagement.

- Example: AI might detect that "Best real estate agent for First-Time Buyers" performs well in search rankings and recommend emphasizing it in content.

Author Note: More hype than reality.

Benefits:

- Makes collecting, organizing, and showcasing reviews effortless.
- Increases engagement by turning reviews into compelling marketing materials.
- Helps attract new leads by reinforcing trust and credibility.
- Improves SEO rankings when social proof is incorporated into website content.

Considerations:

- AI can assist in collecting and formatting reviews, but authenticity matters—never edit testimonials to misrepresent client experiences.
- While automation helps, genuine interaction is key—always thank clients for positive reviews personally.
- Make sure social proof is diverse—highlight testimonials from different types of buyers and sellers to appeal to a broader audience.

AI-powered social proof helps real estate agents build trust, attract new clients, and strengthen their brand effortlessly.

AI for Reputation Management in Real Estate

Status:	**Emerging**
Your priority:	**Very Low**
Hype v. Reality:	**Midpoint**
Effort:	**Medium to Hard**
Examples:	Google Alerts (google.com/alerts) Brand24 (brand24.com) Reputation.com (reputation.com) Yext (yext.com) like Moz Pro (moz.com) Semrush (semrush.com)

Author Note: Having a Google alert for your name (and spelling variations, as needed) is critical, and easy to do. We'd suggest you add alerts for your areas that you farm or track, too. Most of these ideas would apply to businesses that get a lot of reviews. This doesn't apply to most agents unless they run a big team.

How It Works:

Reputation management is about monitoring, protecting, and improving your online presence so potential clients see you as trustworthy and professional. Unlike social proof, which highlights positive feedback, reputation management involves responding to reviews, correcting misinformation, managing online listings, and ensuring your digital brand stays strong. AI-powered tools can track what is being said about you online, suggest responses to reviews, and help control how you appear in search results.

Steps to Do It:

1. **Monitor Your Online Reputation Automatically:**
   - AI-powered tools like Google Alerts (google.com/alerts) and Brand24 (brand24.com) track mentions of your name, brokerage, or brand across the web.
   - AI analyzes reviews and comments to detect patterns, such as frequent concerns about response times or praise for negotiation skills.
2. **Respond to Reviews with AI Assistance:**
   - AI tools like **Reputation.com** (reputation.com) and ChatGPT can suggest professional responses to both positive and negative reviews.
   - Example:
     - **Negative Review:** "The agent took too long to respond."
     - **AI-Suggested Response:** "I appreciate your feedback and apologize for any delays. I strive to provide timely service and will work on improving communication. Thank you for your honesty."
   - AI can auto-flag serious complaints that may require personal follow-up.
3. **Ensure Online Listings Are Accurate and Consistent:**
   - AI tools like Yext (yext.com) scan and update Google Business Profile, Zillow, Realtor.com, and other listings to ensure your contact information, bio, and branding are consistent across platforms.
   - This helps avoid confusion and improves your search engine rankings.
4. **Optimize Your Search Engine Reputation (SEO for Your Name):**
   - AI-powered SEO tools like Moz Pro (moz.com) or Semrush (semrush.com) help control what shows up when someone Googles your name.
   - AI can suggest publishing fresh content (blog posts, videos, interviews) to push down outdated or irrelevant search results.
5. **Manage Your Social Media Presence with AI:**

- AI tools like Hootsuite and Sprout Social analyze engagement and help determine when and how often to post to maintain a strong online presence.
- AI-generated insights can help flag negative trends (e.g., declining engagement, rising negative feedback) so you can adjust your strategy proactively.

Benefits:

- Protects your reputation by addressing negative reviews and misinformation before they spread.
- Saves time by automating responses while keeping communication professional.
- Ensures accuracy across all online platforms, improving brand consistency.
- Helps control search results so your strongest content appears first.

Considerations:

- AI-generated responses should still sound human and professional—always review before posting.
- Ignoring bad reviews is a mistake—AI can help, but personal outreach is sometimes necessary.
- Over-automation can feel impersonal—balance AI with real engagement.

AI-powered reputation management helps real estate agents maintain trust, correct misinformation, and stay in control of their online presence.

Innovative Applications of AI

Automated Follow-Up

Status:	**Emerging**
Your priority:	**Very Low**
Hype v. Reality:	**Midpoint**
Effort:	**Hard**
Examples:	Follow Up Boss (followupboss.com) Real Geeks (realgeeks.com) ActiveCampaign (activecampaign.com) Brevo (brevo.com) Drift (drift.com) ManyChat (manychat.com)

How It Works:

Following up with leads, past clients, and referrals is essential, but it can be time-consuming. AI-powered tools help by automating follow-up emails, texts, and calls, ensuring no lead falls through the cracks. AI can personalize messages based on client behavior, schedule reminders for check-ins, and even generate responses based on previous conversations. This keeps potential buyers and sellers engaged while allowing agents to focus on high-value tasks.

Steps to Do It:

1. **Use an AI-Powered CRM for Follow-Up Tracking:**
   - Platforms like Follow Up Boss (followupboss.com) and Real Geeks (realgeeks.com) track lead interactions and suggest the best time to follow up.
   - AI can analyze which leads are most engaged (e.g., those who opened emails, clicked on listings) and prioritize them.
2. **Automate Email and Text Follow-Ups:**
   - AI-powered tools like ActiveCampaign (activecampaign.com) and Brevo (brevo.com) create personalized email and text sequences for different types of leads.
   - Example: AI-generated follow-up after a listing inquiry:
     - "Hi [First Name], I saw you checked out [property address]. Let me know if you'd like more details or to schedule a tour!"
3. **Schedule AI-Generated Drip Campaigns:**
   - AI can send a sequence of messages over weeks or months to nurture cold leads.
   - Example: A six-email sequence for a buyer lead might include:
     - **Week 1:** "How to Get Pre-Approved for a Mortgage
     - **Week 2:** ""What to Look for in a Home Inspection"
     - **Week 3:** "5 Tips for Negotiating the Best Deal"
4. **Use AI Chatbots for Instant Responses:**
   - AI chatbots like Drift (drift.com) and ManyChat (manychat.com) answer common client questions immediately and capture leads outside of business hours.
   - Example: A chatbot can respond when a visitor asks, "Is this home still available?" and offer to connect them with an agent.
5. **Analyze Follow-Up Effectiveness with AI:**
   - AI tools track which follow-ups get responses and which messages need improvement.
   - Example: AI may suggest sending follow-ups at different times based on past client engagement data.

Benefits:

- Ensures consistent follow-ups without manual effort.

- Saves time by automating lead nurturing and client check-ins.
- Increases conversion rates by keeping leads engaged.
- AI prioritizes hot leads while keeping cold leads warm.

Considerations:

- AI can handle routine follow-ups, but high-value clients still need personal outreach.
- Messages should be customized to avoid feeling overly automated.
- AI is a tool, not a replacement—agents should still build genuine relationships.

AI-powered automated follow-ups help real estate agents stay top-of-mind, convert more leads, and saves valuable time.

Event Planning

Status:	**Emerging**
Your priority:	**Low to Medium**
Hype v. Reality:	**Midpoint**
Effort:	**Medium**
Example:	Eventbrite AI (eventbrite.com) Evite (evite.com) Canva (canva.com) Calendly (calendly.com)

How It Works:

Client appreciation events help real estate agents strengthen relationships, generate referrals, and build brand loyalty. However, planning and executing events can be time-consuming. AI-powered tools can help select event ideas, manage invitations, organize logistics, and even automate follow-ups. Whether hosting a holiday party, homebuyer seminar, or VIP networking event, AI streamlines the entire process.

Author Note: "Streamline" is subjective. We think you'll find it a useful tool, but do keep your expectations in check.

Steps to Do It:

1. **Use AI to Brainstorm and Plan the Event:**
   - AI tools like ChatGPT or Jasper AI can suggest event themes, activities, and locations based on your goals.
   - Example: AI might suggest a "Home Equity Happy Hour" for past clients to discuss market trends and potential sales.

- AI-powered tools like Eventbrite AI (eventbrite.com) help predict event popularity and optimize ticket pricing.

2. **Automate Invitations and RSVPs:**
    - Evite (evite.com) and Canva (canva.com) use AI to generate custom invitations.
    - AI-powered CRM tools like HubSpot or Zoho automate email and text invites and track RSVPs.
    - Example: AI-generated invite message:
        - "Join us for a Client Appreciation Brunch! Enjoy great food and drinks while we celebrate YOU. RSVP here: [Event Link]."
3. **Optimize Event Promotion with AI:**
    - AI tools like Hootsuite and Buffer schedule and optimize social media posts about the event.
    - Example: AI might suggest posting a countdown reminder or a "what to expect" video to boost engagement.
    - Meta (Facebook) Ads AI can analyze which audience is most likely to attend and optimize ad spend accordingly.
4. **Use AI for Event Logistics and Coordination:**
    - AI-powered scheduling tools like **Calendly** help organize guest appointments (e.g., VIP client meetups).
    - Chatbots can answer common questions (e.g., "Where is parking?" or "What time does the event start?").
    - Example: AI can create a run-of-show schedule, ensuring smooth event execution.
5. **Automate Follow-Ups and Lead Nurturing Post-Event:**
    - AI tools like ActiveCampaign and Brevo can send thank-you emails, collect feedback, and nurture leads.
    - Example: AI-generated post-event email:
        - "Thanks for attending our event! We loved seeing you. If you have any real estate questions, let's connect."
    - AI can analyze engagement data (who attended, who clicked the follow-up email) to suggest the best leads to contact personally.

Benefits:

- Saves time by automating invitations, scheduling, and follow-ups.
- Increases event attendance with targeted AI-powered marketing.
- Improves client engagement and builds stronger relationships.
- Identifies which clients are most engaged, helping prioritize follow-ups.

Considerations:

- AI helps organize and optimize, but personal interaction is key at the event.
- AI-generated content should still match your brand's tone and style.

- Not all clients prefer digital invites—consider a mix of email, social media, and direct mail.

AI-powered event planning allows real estate agents to host memorable client appreciation events with less hassle, ensuring strong client relationships and repeat business.

Summary of Concepts

How to Attract Clients with AI

Finding new clients is one of the biggest challenges for real estate agents, but AI is making it easier and more efficient. This chapter explored how AI can streamline lead generation, enhance communication, and improve marketing, allowing agents to focus on building relationships and closing deals.

AI-Powered Content Creation

AI helps real estate agents generate compelling content for blogs, social media, and videos. Tools like ChatGPT, Jasper AI, and Canva assist in writing posts, creating captions, and editing visuals. AI-generated voiceovers and faceless videos allow agents to produce high-quality marketing materials without appearing on camera. With automated content scheduling, agents can maintain a consistent online presence with minimal effort.

AI-Driven Lead Generation and Qualification

AI identifies high-potential leads using predictive analytics and automated scoring. Platforms like Offrs, Likely.ai, and Revaluate analyze online behavior and demographic data to determine which homeowners are most likely to buy or sell. Chatbots and virtual assistants capture leads 24/7 by answering questions and booking appointments automatically. AI-generated email campaigns keep prospects engaged and nurture cold leads until they're ready to act.

Enhancing Client Communication and Relationship Management

AI personalizes client interactions through automated CRM systems like Follow Up Boss and kvCORE. AI-powered social media tools schedule posts and suggest content to increase engagement. Referral marketing benefits from AI-driven reminders to follow up with past clients and request testimonials, ensuring strong relationships and repeat business.

SEO Optimization and Market Analysis

AI improves real estate agents' online visibility by optimizing website content for search engines. AI-driven market analysis tools turn complex data into clear reports, helping agents educate buyers and sellers on current trends. While AI-generated reports save time, agents should verify accuracy before sharing insights with clients.

Social Proof and Reputation Management

AI helps manage online reviews, respond to feedback, and highlight positive client testimonials. Tools like Boast.io and Trustpilot automate review collection, while AI-generated responses ensure prompt and professional replies to online comments. AI can also monitor an agent's online reputation, ensuring accurate and consistent branding across platforms.

Chapter 4 Summary

AI automates follow-up messages, ensuring no lead is forgotten. AI-powered event planning streamlines client appreciation events by handling invitations, scheduling, and post-event follow-ups. These tools help real estate agents strengthen relationships and generate referrals with less manual effort.

AI isn't a replacement for real estate agents—it's a tool that enhances efficiency and effectiveness. By leveraging AI for marketing, lead generation, and client communication, agents can grow their businesses while focusing on what matters most: building lasting relationships.

5. BUYER PROCESS

As a real estate agent, your most valuable asset isn't just your market knowledge or negotiation skills. It is your ability to educate and guide your clients through one of the biggest financial decisions of their lives. Whether you're working with first-time buyers, seasoned investors, or sellers preparing to list, the questions and concerns you hear are often the same.

That's where AI-powered tools and smart client education come in. Instead of answering the same questions repeatedly, you can leverage pre-recorded video content to streamline your communication, build trust, and enhance your client experience.

This book is about how real estate agents can use AI to elevate their business—but it's not just about automation and efficiency. It's about using technology to create a better client experience. And part of that includes offering structured, high-quality educational content that clients can access on their own time.

Buyer Video Library

Buying a home is one of the biggest financial and emotional decisions someone will ever make. In real estate, knowledge is power. The more your client understands about the process, the more confident and prepared they'll feel when making decisions. Whether they are a first-time homebuyer or just need a refresher, these 25 videos are designed to walk your client through every step of the process—from getting pre-qualified for a mortgage to understanding closing costs, handling contract negotiations, and finally getting the keys to their new home.

That's why we've developed this 25-video series—a tool you can share with your buyers to educate them before they ever sit down with you. These videos will help you:

- **Save time** – Instead of explaining the same concepts repeatedly, you can provide a link to a video and focus your in-person time on more complex, high-value discussion.
- **Build trust** – Clients who feel informed are more confident in your expertise, making them more likely to follow your advice.
- **Scale your business** – Educational content helps you serve more clients with less effort, allowing you to focus on higher-value activities like negotiation and strategy.

How to Use This Book and Video Series

This book is structured to help you understand and implement AI in your real estate business, with a focus on practical applications. The video series is one such tool—an AI-enhanced way to create consistent, high-quality client education at scale.

Following is a library of video scripts tuned for first-time buyers. Also, at the end of this book are appendices with all the videos listed. We have included QR codes which will take you to our website, where we have the completed Buyer and Seller videos with voiceover all set to go. You'll also find the script that was used, which you can copy and personalize for your brand. You might have specific changes you need to make for your local market. For example, if well or septic systems are common where you work, you'd want to add that to your inspection video.

Here's how you can use this series:

- **Before Buyer Consultations** – Send the first few videos to new clients so they come to your meeting prepared.
- **During the Transaction** – Share videos on inspections, appraisals, and closing steps to guide clients through the process.
- **After Closing** – Use the final videos to help clients transition into homeownership and stay engaged for future referrals.

Technology should enhance your role as a real estate agent, not replace it. By using AI-powered solutions and pre-built educational content, you can elevate your brand, save time, and provide an even better experience for your clients.

Buyer Video Library

Seller Video Library

Buyer Onboarding and Preparation

VIDEO 1 – New Buyer Client Onboarding Script

Opening:
If you're watching this, chances are you're thinking about buying your first home—or maybe you've already started the process and want to make sure you're on the right track. First off, congratulations! Buying a home is one of the most exciting milestones in life, and I'm here to help make it as smooth and stress-free as possible.

What to Expect:
Buying a home may feel overwhelming, but don't worry. In this video, I'll walk you through the key steps you'll take as a first-time homebuyer. I'll cover how we'll work together, what you need to do upfront, and how we'll make sure you find the right home

Step 1: Understanding Your Goals
The first thing we'll do is sit down for a buyer consultation. This is a no-pressure conversation where we talk about what's important to you—your must-haves, your budget, your timeline, and any concerns you have. My job is to understand your goals so we can tailor the process to fit you.

Step 2: Getting Pre-Qualified for a Mortgage
One of the most important first steps is getting pre-qualified for a mortgage. This helps us understand what price range we should be looking at and makes you a stronger buyer when we submit an offer. I can connect you with a trusted lender if you don't already have one, and they'll help you determine what you can afford, what loan options you have, and what your estimated monthly payments will look like.

Step 3: Understanding the Market and The Search Process
Once we have your pre-qualification in hand, we can start looking at homes! I'll set up personalized alerts so you see listings that match your criteria as soon as they hit the market. We'll also discuss what to expect in today's market—whether it's a buyer's market or a seller's market—so you're prepared for the competitive landscape.

Step 4: Touring Homes and Making a Decision
As we tour homes, I'll help you evaluate the pros and cons of each property. I'll also guide you on what to look for beyond just aesthetics—things like resale value, potential red flags, and whether a home is priced fairly. When you find the right home, we'll work together on making a strong, competitive offer.

Step 5: The Offer, Negotiation, and Going Under Contract
Once we submit an offer, the seller can accept, reject, or negotiate. I'll walk you through the negotiation process, ensuring we get you the best deal possible while protecting your interests. If your offer is accepted, we'll go under contract; this phase includes inspections, appraisal, and final mortgage approval.

Step 6: Closing the Deal and Moving In
During the closing process, we'll handle final paperwork, conduct a final walkthrough, and make sure everything is set for a smooth closing day. On closing day, you'll sign the final documents, get the keys, and officially become a homeowner!

Final Thoughts:
Buying your first home is an exciting journey, and I'm here to guide you every step of the way. If you have any questions, don't hesitate to reach out.

VIDEO 2 – Preparing for a Buyer Consultation

Opening:
If you're watching this, you've probably scheduled—or are thinking about scheduling—a buyer consultation with your real estate agent. First, great decision! This is one of the most important steps in the home-buying process because it helps set expectations, answer key questions, and make sure we're on the same page.

What is a Buyer Consultation?
A buyer consultation is a dedicated meeting between you and your real estate agent where we discuss your home-buying goals, review the process, and ensure you feel confident moving forward. It's designed to eliminate surprises and make the journey as smooth as possible.

Step 1: What to Bring and How to Prepare
To get the most out of our consultation, here are a few things you can do to prepare:

- Think about your must-haves and deal-breakers—the features that matter most to you in a home.
- Consider your budget and financing options—if you're not yet pre-qualified for a mortgage, I can connect you with some great lenders to choose from.
- Make a list of any questions or concerns you have about the buying process.

Step 2: Understanding the Home Buying Process
During our meeting, I'll walk you through each phase of the home-buying journey, including:

1. **Getting pre-qualified** for a mortgage.
2. **Searching for the right home** and setting up custom alerts.
3. **Touring properties** and knowing what to look for beyond aesthetics.
4. **Making an offer**—how we structure a competitive bid.
5. **The contract phase**—what to expect with inspections, appraisals, and contingencies.
6. **Closing day!** Finalizing the deal and getting your keys.

Step 3: What Happens After the Consultation?
After our meeting, you'll have a clear action plan. If you're ready to start searching, we'll set up custom listing alerts and schedule showings. If you need to get pre-qualified, I'll introduce you to a few lenders so you know your budget before we start touring homes.

Final Thoughts:
My goal is to make sure you feel confident and informed every step of the way. If you haven't already scheduled your buyer consultation, let's set that up today! I look forward to helping you find the perfect home.

VIDEO 3 – Why Get Pre-Qualified for a Mortgage

Opening:

If you're thinking about buying a home, one of the most important first steps is getting pre-qualified for a mortgage. You might be wondering why this is necessary. In this video, I'll explain why pre-qualification is crucial, how to prepare for your lender meeting, and what to do afterward.

Step 1: Why Pre-Qualification Matters
Getting pre-qualified helps you understand what you can afford, gives you a competitive edge when making offers, and shows sellers that you're a serious buyer. In today's market, homes can move fast, and having your pre-qualification in hand makes your offer much stronger. Plus, it prevents you from falling in love with a home that might be outside your budget.

Step 2: How to Prepare for Your Lender Meeting
Before meeting with a lender, gather the following documents:

- **Proof of income** – Recent pay stubs, W-2s, or tax returns if you're self-employed.
- **Proof of assets** – Bank statements, investment accounts, and savings.
- **Credit information** – Lenders will check your credit score, so review your credit report beforehand.

- **Employment verification** – A letter from your employer or recent work history.
- **Debt obligations** – Any outstanding loans, credit cards, or other financial commitments.

Having these documents ready will speed up the process and give you a more accurate picture of what you can afford.

Step 3: Good Questions to Ask Your Lender

When you meet with your lender, here are some key questions to ask:

- What types of loans do I qualify for, and what are the differences?
- What's the interest rate, and is it fixed or adjustable?
- What will my monthly payments look like, including taxes and insurance?
- Are there any down payment assistance programs available?
- What are the closing costs, and can any be negotiated?

Step 4: What to Do After the Meeting

Once you're pre-qualified, stick to smart financial habits. Avoid making large purchases, opening new credit lines, or changing jobs before closing on your home. Lenders will re-check your financial situation before final approval, so keeping everything stable is key.

Final Thoughts:

Getting pre-qualified is an essential step in your home-buying journey. It sets you up for success, gives you clarity on your budget, and strengthens your offer. If you have any questions or need lender recommendations, reach out—I'm happy to help! Let's get you on the path to homeownership.

VIDEO 4 – Understanding Buyer Closing Costs

Opening:

Whether you're buying or selling a home, **closing costs** are an important part of the transaction that many people don't fully understand until the last minute. In this video, I'll break down what closing costs are, how much you should expect to pay, and how to prepare so there are no surprises.

Step 1: What Are Closing Costs?

Closing costs are the fees and expenses that both buyers and sellers pay at the closing table to finalize the real estate transaction. These fees cover things like loan processing, legal documentation, and taxes. The total amount varies based on the property price, location, and loan type.

Step 2: Closing Costs for Buyers
As a buyer, your closing costs typically range from 2% to 5% of the home's purchase price. These include:

- **Loan Origination Fees** – Charged by your lender for processing the mortgage.
- **Appraisal and Credit Report Fees** – Covers the cost of the home appraisal and checking your credit.
- **Title Insurance and Title Search** – Protects you and the lender from ownership disputes.
- **Escrow Fees** – Paid to the title company handling the transaction.
- **Prepaid Property Taxes and Homeowners Insurance** – Usually required to be paid in advance.
- **Recording Fees** – Charged by government entities to register the home in your name.
- **HOA Fees** – If you are buying into a community with an HOA, you may be required to pay two months of HOA dues upfront, HOA reserve fees, and or record change fees.
- **Commissions** – Paid to your real estate agent for handling the transaction.

Step 4: How to Prepare for Closing Costs
To avoid last-minute surprises:

- **Get a loan estimate** from your lender early in the process so you know what to expect.
- **Ask your real estate agent or lender** about ways to reduce costs, such as negotiating seller concessions or lender credits.
- **Check in with your real estate agent** approximately a week before closing to see if a preliminary settlement statement is available.

Final Thoughts:
Closing costs are a key part of buying or selling a home and understanding them in advance helps you budget and plan accordingly. If you have any questions about your specific closing costs or need help estimating them, reach out—I'm happy to help!

VIDEO 5 – Your Dream First Home vs. What You Can Afford

Opening:

If you're a first-time homebuyer, you've probably imagined your perfect dream home. Maybe it has a huge kitchen, a big backyard, or that perfect open floor plan. But, when it comes time to start looking, reality sets in—your budget might not match your wish list. In this video, I'll help you reconcile what you want in your first home vs. what you can realistically afford while still making a smart investment.

Step 1: Defining Your Priorities

The first step is to separate your must-haves from your nice-to-haves. Ask yourself:

- What features do I need to be comfortable in my first home?
- What things can I live without for now?
- What aspects are non-negotiable? (Location, number of bedrooms, etc.)

Your first home doesn't have to be your forever home, but it should meet your most important needs.

Step 2: Understanding Your Budget

Next, let's talk numbers. Just because you're pre-qualified for a certain amount doesn't mean you should spend every dollar of that. Consider these factors:

- Monthly mortgage payment (including taxes & insurance)
- Maintenance costs—older homes might need more repairs.
- Lifestyle expenses—do you still want room in your budget for travel, dining out, or hobbies?

Buying within your means ensures financial stability and reduces stress.

Step 3: Being Open to Trade-Offs

Many first-time buyers have to make trade-offs. Here are a few common ones:

- **Size vs. Location** – A smaller home in a great neighborhood vs. a bigger home further away.
- **Move-in Ready vs. Fixer-Upper** – A newer home that's turn-key vs. an older home with renovation potential.
- **Amenities vs. Affordability** – Extra features like a pool or big backyard vs. staying within budget.

Step 4: Thinking Long-Term

Your first home should also be a smart investment. Consider:

- Will this home appreciate in value?
- Is it in a good school district or a growing area?
- Can you see yourself staying for at least 3-5 years?

Thinking ahead helps ensure that when you're ready to upgrade, you'll have equity to put toward your next home.

Final Thoughts:

Your first home doesn't have to be your dream home—it just has to be the **right** home for **now**. By balancing your wish list with a realistic budget, you'll find a home that makes sense financially and still feels like a great place to live. If you need help prioritizing your search, let's set up a time to talk! I'd love to guide you through this exciting process.

The Home Search and Making an Offer

VIDEO 6 – Why Buying is Better Than Renting for Most

Opening:
If you're a first-time homebuyer, you've probably asked yourself: 'Should I buy a home, or should I keep renting?' This is a huge decision, and while renting can be the right choice in some situations, for most people, **buying is the better long-term financial move.** In this video, I'll guide you through the key reasons why buying often makes more sense than renting—and how it can help build your financial future.

Step 1: Building Equity Instead of Paying Rent
When you rent, you're essentially paying someone else's mortgage. Every month, your rent check goes to your landlord, and you build no long-term value for yourself. When you own a home, your mortgage payments help you build equity, which means you're increasing your financial stake in your property. Over time, as home values increase, your investment grows.

Step 2: Stability & Predictable Costs
Rent prices can go up every year—sometimes dramatically—making it hard to budget long-term. When you buy a home with a fixed-rate mortgage, your monthly payment stays consistent, giving you greater financial stability. This allows you to plan for the future without worrying about sudden rent hikes.

Step 3: Tax Benefits of Homeownership
Homeownership comes with significant tax advantages that renters don't get. You may be able to deduct mortgage interest, property taxes, and even some closing costs, which can reduce your overall tax bill. Be sure to consult with a tax professional to understand the benefits available to you.

Step 4: Freedom to Make It Your Own
When you rent, you're often limited in what you can change about your living space. Want to paint the walls? Upgrade appliances? Knock down a wall? That's usually not an option as a renter. As a homeowner, you have complete control over customizing your space to fit your style and needs.

Step 5: Long-Term Financial Growth
Real estate is one of the most powerful wealth-building tools available. Over time, home values tend to rise, allowing you to sell later at a profit. That profit can help you upgrade to a larger home, invest in property, or fund other financial goals.

Instead of throwing money away on rent, your mortgage payments work toward your future wealth.

When Renting Might Be the Better Choice
Now, buying isn't for everyone. If you're planning to move in the next year or two, if your job situation is unstable, or if you're not ready for the responsibilities of homeownership, renting might be the better option for now. But if you're planning to stay in one place for a few years and want to build long-term wealth, buying is likely the smarter move.

Final Thoughts:
Renting might seem like the easier option, but in most cases, owning a home is a smarter financial decision that offers stability, tax benefits, and long-term wealth-building potential. If you're thinking about buying but aren't sure where to start, let's talk! I'd love to help you explore your options and take the next step toward homeownership.

VIDEO 7 – Elements of the Home Search

Opening:
If you're starting the exciting journey of finding a home, you might be wondering—what exactly does the home search process look like? And how can you work effectively with your real estate agent to find the perfect place? In this video, I'll explain the key steps of the home search process, set expectations, and share tips to make the experience smooth and successful.

Step 1: Defining Your Needs and Wants
Before we start looking at homes, the first step is to define what you truly need vs. what you want. Consider:

- Number of bedrooms and bathrooms
- Ideal location and commute time
- Style and type of home (single-family, condo, townhouse, etc.)
- Must-have features vs. nice-to-have features

This helps us narrow down your search so we're only looking at homes that match your lifestyle and priorities.

Step 2: Getting Your Search Set Up
Once we have your criteria, I'll set up a custom MLS search so you receive alerts when new listings hit the market. We'll also go over other sources, like off-market opportunities, new construction homes, and open houses to maximize your options.

Step 3: Touring Homes and Evaluating Options
Now comes the fun part—touring homes! As we visit properties, I encourage you to:

- Take notes and compare homes objectively.
- Look beyond cosmetic details and focus on layout, structure, and long-term potential.
- Ask questions—I'm here to help you evaluate each home with a professional eye.

Step 4: How to Decide Which Home is Right for You
When you find a home you love, consider:

- Does it meet your non-negotiable needs?
- Can you see yourself living there for at least a few years?
- Is it within your budget, including maintenance and future expenses?

If the answer is yes, then we'll talk about making an offer.

Step 5: Working Effectively with Your Real Estate Agent
Your real estate agent is your guide through this process. To get the most out of our partnership:

- Be open about your feedback after showings so we can refine the search.
- Be responsive—homes move fast, and communication is key.
- Trust the process—we'll make sure you're making a well-informed decision.

Final Thoughts:
The home search process should be exciting, not stressful. By knowing what to expect and working closely together, we'll find the right home for you. If you're ready to start your search, let's set up a time to talk! I'll help you navigate this journey.

VIDEO 8 – What It's Like to Go Under Contract

Opening:
Congratulations—your offer has been accepted, and you're officially under contract on your first home! This is an exciting milestone, but I know it can also feel overwhelming. In this video, I'll guide you through what happens next, your responsibilities as a buyer, and what to expect in the coming weeks.

Step 1: Understanding the Key Deadlines
Once you're under contract, we'll be working within specific timeframes, including:

- **Earnest Money Deposit (EMD)** – This needs to be submitted within the first few days to show your commitment to the purchase.
- **Home Inspection Period** – This is your opportunity to thoroughly inspect the property and negotiate any necessary repairs.
- **Appraisal and Loan Approval** – The lender will assess the home's value and finalize your mortgage.
- **Title Review and HOA Documents (if applicable)** – You'll have time to review these to ensure there are no hidden concerns.

Step 2: The Home Inspection
Your home inspection is one of the most important steps. You'll want inspector will evaluate the property's condition, including:

- Structural integrity
- Electrical and plumbing systems
- Roof and HVAC condition

If issues arise, we can negotiate with the seller for repairs or credits before moving forward.

Step 3: The Appraisal Process
Your lender will order an appraisal to determine the home's value. If the appraisal comes in **at or above** the contract price, great! If it comes in low, we'll discuss options such as renegotiating with the seller or covering the difference."

Step 4: Preparing for Closing
As we get closer to closing day:

- Stay in constant communication with your lender to ensure everything is on track.
- Avoid making major financial changes—no big purchases or new credit lines!
- Secure homeowner's insurance and arrange for utilities to be transferred.

Step 5: Final Walkthrough and Closing Day
A day or two before closing, we'll do a final walkthrough to ensure the home is in the agreed-upon condition. Then on closing day, you'll sign the final paperwork, submit your down payment and closing costs, and **get the keys to your new home!**

Final Thoughts:
Being under contract is an exciting but busy time, and I'll be with you every step of the way. If you have any questions about what's ahead, don't hesitate to reach out. I can't wait to celebrate with you at closing!

VIDEO 9 – Common Contract Pitfalls and How to Avoid Them

Opening:
Real estate contracts can be complex, and even small mistakes can cause major delays—or worse, derail the entire transaction. In this video, I'll explain some of the most common contract pitfalls and how you can avoid them to ensure a smooth buying or selling experience.

Step 1: Missing Key Deadlines
Contracts come with strict timelines, including deadlines for inspections, financing approval, and closing. If any of these are missed, the deal could fall apart. To stay on track. Work closely with your lender and real estate agent to ensure financing processes and inspections happen on time.

Step 2: Contingency Confusion
Contingencies are conditions that must be met for the contract to move forward, such as:

- **Financing Contingency** – Protects buyers if their loan falls through.
- **Inspection Contingency** – Allows buyers to negotiate repairs or back out if major issues arise.
- **Appraisal Contingency** – Ensures the home's value meets or exceeds the agreed-upon price.

To avoid issues, make sure you fully understand what contingencies are in place and what actions are required to meet them.

Step 3: Incomplete or Incorrect Paperwork
Even small errors on the contract—like a misspelled name or wrong closing date—can cause delays. Always:

- Double-check all details before signing.
- Ensure all required signatures and initials are in place.
- Review the final contract with your real estate agent to catch any mistakes.

Step 4: Financing Issues
Many deals fall apart due to financing problems, such as a low appraisal or last-minute loan denial. To prevent this:

- Get **fully pre-approved**, not just pre-qualified.
- Avoid making big purchases or changing jobs before closing.

Step 5: Misunderstanding Earnest Money
Earnest money is a deposit buyers put down to show they're serious about the purchase. If the buyer backs out without a valid contingency, they could lose this deposit. To avoid problems, make sure you know the conditions under which you can get your earnest money back.

Final Thoughts:
Real estate contracts are legally binding, and avoiding these common pitfalls can save you time, money, and stress. If you have any questions or need guidance navigating a contract, reach out—I'm here to help ensure a smooth and successful transaction!

VIDEO 10 – Negotiating the Contract

Opening:

You've found a home you love, and now it's time for one of the most important steps—negotiating the contract. This part of the process can be exciting, but it's also where having a strong strategy matters. In this video, I'll talk about what to expect, your role in the negotiation process, and how we work together to get the best deal possible.

Step 1: Understanding the Key Negotiation Points
A real estate contract isn't just about the purchase price. Here are a few key areas we may negotiate:

- **Price** – Based on market conditions, comparable sales, and the home's condition.
- **Closing Costs** – Sometimes, we can ask the seller to contribute towards your closing costs.
- **Inspection Repairs** – If issues come up in the home inspection, we can negotiate for repairs or a credit.
- **Closing Date and Possession Terms** – Flexibility here can sometimes be a negotiation tool.

Step 2: Your Responsibilities as a Buyer
Negotiation isn't just about what we ask for—it's also about being prepared. Here's what you can do to strengthen our position:

- Be clear on your budget and what you're willing to compromise on.
- Stay responsive—Negotiations move quickly, and delays can weaken our position.
- Trust the strategy—Sometimes, counteroffers and small concessions help seal the deal.

Step 3: Handling Counteroffers and Moving Forward
It's common for sellers to counter our offer rather than accept the first one. If that happens, we'll discuss options and decide on the next best move—whether it's accepting, countering again, or walking away if the terms don't work for you.

Step 4: When to Walk Away
Sometimes, despite our best efforts, a deal isn't the right fit. If the seller isn't willing

to meet reasonable terms or the home has unexpected issues, it's okay to walk away and find something better suited for you.

Final Thoughts:
Negotiating a home purchase is part art, part strategy, and part timing. My job is to guide you through the process, advocate for your best interests, and make sure you feel confident every step of the way. If you have any questions about the negotiation process, let's chat!

VIDEO 11 – How to Handle Multiple Offers

Opening:
If you're buying or selling a home in a competitive market, you're likely to run into multiple offers. As a buyer, it can be stressful to compete with others for your dream home. In this video, I'll break down the best strategies when dealing with multiple offers.

Strategies for Buyers to Win in a Multiple-Offer Situation
If you're a buyer competing in a multiple-offer situation, making your offer stand out is key. Here are some strategies to increase your chances of winning:

- **Get Pre-Approved, Not Just Pre-Qualified** – A pre-approval letter shows sellers you have verified finances and are a serious buyer.
- **Increase Your Earnest Money Deposit** – A higher deposit signals commitment and reassures the seller you're serious.
- **Offer Above Asking Price (When Appropriate)** – If homes are selling quickly, consider offering slightly above asking—within reason and within your budget.
- **Use an Escalation Clause** – This allows you to automatically increase your offer up to a set limit if competing offers come in higher.
- **Be Flexible on Contingencies and Closing Timelines** – Fewer contingencies, such as a waived home warranty or a flexible closing date, can make your offer more attractive.
- **Personalize Your Offer** – Some buyers choose to include a letter explaining why they love the home. Be mindful that some markets discourage these letters due to fair housing concerns.

Final Thoughts:
Whether you're a buyer or seller, handling multiple offers strategically can make a big difference. Buyers should focus on making their offer as strong as possible while staying within their budget.

VIDEO 12 – What Happens If the Deal Falls Apart? (How to Handle a Canceled Transaction and What to Do Next)

Opening:

Buying a home is an exciting journey, but sometimes things don't go as planned. Deals can fall apart for a variety of reasons—maybe the financing didn't go through, an inspection revealed major issues, or the seller had second thoughts. Whatever the reason, I know how frustrating and disappointing this can be. But don't worry! In this video, I'll walk you through why deals fall apart, what to do next, and how to bounce back stronger so you can successfully find the right home.

Step 1: Common Reasons a Deal Falls Through

Understanding why deals fall apart can help you prepare and possibly avoid issues in the future. Here are some of the most common reasons:

- **Financing Issues** – The buyer's loan wasn't approved due to a credit issue, job change, or lender problem.
- **Home Inspection Problems** – Major defects like foundation issues, roof damage, or safety concerns scared the buyer away.
- **Low Appraisal** – The home appraised for less than the agreed purchase price, and the buyer and seller couldn't agree on a solution.
- **Title Issues** – The home's title had legal disputes, unpaid liens, or ownership concerns that weren't resolved.
- **Buyer or Seller Got Cold Feet** – Sometimes, emotions take over, and either party decides to back out.
- **Missed Deadlines or Contract Issues** – Key contract deadlines were missed, or the seller refused to negotiate necessary repairs.

Step 2: What to Do Immediately If Your Deal Falls Apart

If your deal just collapsed, take a deep breath – it's disappointing, but you still have options! Here's what to do next:

- **Talk to Your Real Estate Agent and Lender** – Understand exactly why the deal fell through and whether anything can be done to revive it.
- **If It's a Financing Issue, Find Out Why** – Work with your lender to correct any issues and get fully pre-approved before making another offer.
- **If It's an Inspection Issue, Consider Alternative Solutions** – Some problems can be fixed through negotiation, credits, or repairs instead of walking away.
- **If It's a Low Appraisal, Explore Options** – You can request a second appraisal, negotiate a lower price, or cover the difference if possible.
- **Review Your Earnest Money Status** – Depending on the reason for the cancellation, you may be able to recover your deposit or renegotiate your terms.

Step 3: How to Protect Yourself in the Future
To avoid another canceled deal, take these proactive steps:

- **Get Fully Pre-Approved (Not Just Pre-Qualified)** – This ensures financing won't be a last-minute issue.
- **Understand Your Contract Contingencies** – Know your rights when it comes to inspections, appraisals, and financing deadlines.
- **Ask for a Seller's Disclosure** – Before making an offer, review any potential home issues upfront to avoid surprises.
- **Be Realistic in Your Expectations** – No home is perfect; be prepared to compromise on small issues while focusing on major deal-breakers.
- **Work with an Experienced Real Estate Team** – Having a knowledgeable agent and lender can help navigate challenges and keep deals on track.

Step 4: Moving Forward with Confidence
It's completely normal to feel frustrated or discouraged after a deal falls apart. But remember: this wasn't the right home, and that's okay! There's another great home out there waiting for you. Here's how to keep moving forward:

- **Stay Positive & Keep Looking** – The right home will come along, and now you have even more experience to make a stronger offer next time.
- **Refine Your Search** – Based on what went wrong, adjust your criteria and strategy to avoid similar pitfalls.
- **Be Ready to Act Quickly** – Sometimes, homes that fall out of contract come back on the market—if you loved the house, we might be able to jump back in.
- **Lean on Your Support Team** – Your real estate agent, lender, and support network are here to help you succeed!

Final Thoughts:
A canceled deal isn't the end of your home-buying journey—it's just a detour. The important thing is to learn from the experience, adjust your approach, and keep going. Many buyers go under contract multiple times before closing on the right home, and when you finally get those keys in your hand, it will all be worth it! If you have any questions, need advice, or want to get back out there and find the perfect home, reach out—I'm here to help you every step of the way.

Due Diligence and Financing

VIDEO 13 – What is a Home Inspection? Questions to Ask

Opening:

If you're under contract on a home, one of the most crucial steps is the home inspection. This is your chance to take a deep dive into the condition of the property before finalizing the purchase. In this video, I'll explain what to expect, the key questions to ask, and best practices to ensure you're making a well-informed decision.

Step 1: What is a Home Inspection?
A home inspection is a thorough evaluation of the property's condition conducted by a licensed professional. The inspector will check:

- Structural components like the foundation and roof
- Electrical and plumbing systems
- HVAC system and appliances
- Overall safety concerns and potential future repairs

Step 2: Questions to Ask Your Home Inspector
While the inspector will provide a detailed report, it's important to ask questions during the process, such as:

- Are there any major structural issues I should be concerned about?
- How old are the roof, HVAC system, and water heater?
- Are there signs of water damage or mold?
- What repairs should be prioritized immediately vs. those that can wait?
- Are there any code violations or safety hazards?

Step 3: Your Responsibilities as a Buyer
To make the most of your home inspection:

- Attend the inspection if possible, so you can see issues firsthand.
- Review the report thoroughly and discuss any concerns with me.
- Decide if you want to negotiate repairs or credits before moving forward.

Step 4: What Does the Real Estate Agent Do?
As your real estate agent, my role is to:

- Coordinate the inspection and recommend trusted professionals.
- Help understand which issues may be deal-breakers.
- Negotiate repairs or credits with the seller based on the report.

Final Thoughts:
A home inspection helps ensure you're making a smart investment and aren't hit with unexpected surprises after moving in. If you have any questions or need recommendations for a great home inspector, let's chat! I'm here to help guide you every step of the way.

VIDEO 14 – Reviewing the Title Policy: Questions to Ask

Opening:

When you're buying a home, one of the most important but often overlooked steps is reviewing the title policy. This document ensures that you have clear ownership of the property and protects you from legal issues down the road. In this video, I'll guide you through what a title policy is, why it matters, and the key questions to ask when reviewing yours.

Step 1: What is a Title Policy?

A title policy is a form of insurance that protects you and your lender from any legal claims or disputes regarding ownership of the property. This can include:

- Liens or unpaid debts attached to the home
- Errors in public records
- Ownership disputes or missing heirs
- Fraud or forgery in past transactions

Step 2: Questions to Ask When Reviewing Your Title Policy

Before finalizing your home purchase, review the title policy carefully and ask:

- Are there any existing liens or encumbrances on the property?
- Does the policy include coverage for boundary disputes?
- Are there any restrictions or easements that could affect future use of the property?
- Is there lender's title insurance vs. owner's title insurance—and do I need both?
- What exclusions or limitations does the policy contain?

Step 3: Your Responsibilities as a Buyer

To ensure you have a clean title:

- Carefully review the preliminary title report provided by the title company.
- Ask your title company or attorney to clarify anything that seems unclear.
- Work with your real estate agent to resolve any outstanding title issues before closing.

Step 4: What Does the Real Estate Agent Do?

As your real estate agent, I'll:

- Coordinate with the title company to ensure the property has a clear title.
- Help identify and address potential red flags in the title report.
- Ensure all title issues are resolved before closing day.

Final Thoughts:

A clear title is essential for a smooth home purchase and long-term peace of mind. If you have any questions about your title policy or need help reviewing it, let's chat! I'm here to help make sure everything is in order before you close on your new home.

VIDEO 15 – Reviewing HOA Documents: Questions to Ask

Opening:

If you're buying a home in a neighborhood with a Homeowners Association (HOA), reviewing the HOA documents is a crucial step before closing. These documents outline the rules, fees, and financial health of the community, which can impact your homeownership experience. In this video, we'll discuss what to look for, key questions to ask, and best practices to ensure there are no surprises down the road.

Step 1: What's Included in HOA Documents?

When you receive your HOA documents, you'll typically find:

- **Covenants, Conditions, and Restrictions (CC&Rs)** – The community rules you must follow.
- **Bylaws** – How the HOA is governed and how decisions are made.
- **Financial Statements and Budget** – Details about the HOA's financial health.
- **Reserve Study** – A report showing how well the HOA is prepared for major repairs and maintenance.
- **Meeting Minutes** – Insights into current community issues and potential future changes.

Step 2: Key Questions to Ask When Reviewing HOA Documents

Before you commit, take time to review the documents and ask:

- What are the monthly dues and what do they cover?
- Are there special assessments planned in the near future?
- Are there restrictions on renting the home if I decide to lease it later?
- What are the rules on exterior modifications like painting or landscaping?
- Are there any ongoing legal disputes or financial concerns with the HOA?

Step 3: Your Responsibilities as a Buyer

To ensure you're making an informed decision:

- Read the CC&Rs carefully to understand all restrictions.
- Check for upcoming assessments or financial shortfalls in the HOA budget.
- Talk to current residents if possible, to get a feel for the community and HOA management.

Step 4: What Does the Real Estate Agent Do?
As your real estate agent, I'll:

- Ensure you receive all necessary HOA documents in a timely manner.
- Help identify any red flags in the financials or restrictions.
- Discuss potential concerns and guide you in making a confident decision.

Final Thoughts:
The HOA can have a big impact on your lifestyle and finances, so reviewing these documents thoroughly is essential. If you have any questions or need help interpreting your HOA documents, let's chat! I'm here to help make sure there are no surprises before you close on your new home.

VIDEO 16 – Understanding Home Warranties

Opening:
As a homebuyer, you might hear about home warranties and wonder if they're worth the cost. What do they actually cover? How are they different from homeowners insurance? And should you get one?

Step 1: What is a Home Warranty?
A home warranty is a service contract that covers repairs or replacements of major home systems and appliances due to normal wear and tear. It is different from homeowners insurance, which protects against damage from disasters like fire, flooding, or theft. Most home warranties last for one year, with the option to renew.

Step 2: What Does a Home Warranty Cover?
Most standard home warranties include coverage for:

- **Major Appliances** – Refrigerator, oven, dishwasher, washer, dryer, and garbage disposal.
- **Heating and Cooling Systems** – Furnace, air conditioning, and heat pump.
- **Plumbing and Electrical** – Water heater, plumbing leaks, and electrical systems.
- **Optional Add-Ons** – Some plans offer additional coverage for pool equipment, septic systems, and well pumps.

However, home warranties do **not** typically cover:

- Pre-existing issues before the policy starts.
- Structural damage such as roof or foundation issues.
- Cosmetic repairs.
- Anything not explicitly listed in the contract.

Step 3: Are Home Warranties Worth It for Buyers?

A home warranty can be a smart investment in some cases but unnecessary in others. Consider these factors:

When a Home Warranty Might Be a Good Idea:

- You're buying an older home with aging systems and appliances.
- You want financial protection against unexpected repair costs.
- The seller provides a home warranty as part of the purchase agreement.

When a Home Warranty Might Not Be Worth It:

- You're buying a brand-new home that already has manufacturer warranties.
- You're comfortable handling minor repairs yourself.
- The policy has too many exclusions or high service fees.

Tip: Always compare different home warranty providers—some offer better coverage.

Final Thoughts:
Home warranties can offer peace of mind, especially for first-time buyers who don't want to deal with costly repairs right after moving in. But they aren't always necessary. Carefully review the coverage and costs to see if it fits your needs. If you have questions about home warranties or need help choosing one, reach out—I'd be happy to help!

VIDEO 17 – Real Estate Tax Benefits and Consideration

Opening:
Did you know that buying a home comes with major tax benefits? Many first-time buyers don't realize how much they can save just by being a homeowner. In this video, I'll break down the key tax advantages, what deductions you may qualify for, and what you should know before tax season.

Step 1: Mortgage Interest Deduction
The mortgage interest deduction is one of the biggest tax perks for homeowners. You can deduct the interest paid on your mortgage loan, up to $750,000 in loan balance for most people. This deduction reduces your taxable income, meaning you owe less in taxes.

Tip: Check your **Form 1098**, provided by your lender, to see the total amount of interest you paid during the year.

Step 2: Property Tax Deduction
Homeowners can deduct up to $10,000 in property taxes per year ($5,000 if married and filing separately). This deduction helps lower your taxable income, making it another great financial advantage of homeownership.

Step 3: Mortgage Insurance Deduction
If you put down less than 20% when buying a home, you may be paying private mortgage insurance (PMI). In some cases, PMI payments can be tax-deductible. However, eligibility depends on your income and tax laws, so check with a tax professional to see if you qualify.

Step 4: Capital Gains Exclusion (When You Sell)
When selling your home, you may qualify for a tax break on your profits. If you've lived in the home for at least two of the last five years, you can exclude up to $250,000 in profit from capital gains taxes ($500,000 for married couples). This means you keep more of your home's appreciation when you sell.

Tip: Keep records of major home improvements, as these can help reduce taxable gains when selling.

Step 5: Energy Efficiency Tax Credits
Making energy-efficient upgrades to your home can also lead to tax benefits. Federal tax credits may be available for improvements like installing solar panels, energy-efficient windows, or smart thermostats. These credits directly reduce the amount of tax you owe, rather than just lowering taxable income.

Final Thoughts:
Owning a home isn't just about building equity—it can also mean major tax savings. Whether it's mortgage interest, property taxes, or energy-efficient upgrades, these benefits help put money back in your pocket. If you have questions or need recommendations for a great tax advisor, reach out—I'd love to help!

VIDEO 18 – Understanding the Home Appraisal Process

Opening:
The home appraisal. determines the fair market value of the property and ensures the lender is making a sound investment. In this video, I'll talk about what the appraisal process looks like, key questions to ask, and best practices to make sure everything goes smoothly.

Step 1: What is a Home Appraisal?
A home appraisal is an independent evaluation conducted by a licensed appraiser to assess the fair market value of the home. The appraiser considers factors such as:

- The size, condition, and features of the home
- Recent comparable sales in the area (known as 'comps')
- The local real estate market trends

The lender requires this to confirm that the home is worth the agreed-upon purchase price.

Step 2: What Happens During the Appraisal?
Here's what to expect:

- The appraiser visits the home to inspect its condition and features.
- They analyze recent comparable sales in the neighborhood.
- A final report is issued with the appraised value.

Step 3: Questions to Ask About the Appraisal
While the appraisal process is straightforward, here are key questions you might want to ask:

- How does the appraiser determine the home's value?
- What happens if the appraisal comes in **below** the purchase price?
- Can I challenge an appraisal if I believe it's inaccurate?
- How long does the appraisal process take?

Step 4: Your Responsibilities as a Buyer
To ensure a smooth appraisal process:

- Work with your lender to schedule the appraisal promptly.
- Understand that if the appraisal comes in low, we may need to renegotiate with the seller or explore other options.
- Stay in communication with your real estate agent and lender to keep the process moving forward.

Step 5: What Does the real estate agent Do?
As your real estate agent, I will:

- Ensure the appraiser has access to relevant property details and comps.
- Advocate for an accurate valuation by providing supporting market data if needed.
- Help navigate next steps if the appraisal comes in low.

Final Thoughts:
The home appraisal is a key step in protecting your investment and ensuring you're paying a fair price. If you have any questions about the process or need guidance, let's talk! I'm here to help you every step of the way.

VIDEO 19 – What Happens If the Appraisal Comes in Low?

Opening:
You found the perfect home, got an accepted offer, and now you're one step closer to closing. But then the appraisal comes in lower than the purchase price. Now what? Don't panic. In this video, I'll explain why appraisals come in low, what it means for your deal, and how to keep the transaction on track.

Step 1: Why Do Low Appraisals Happen?
An appraisal is an independent evaluation of a home's value, conducted by a licensed appraiser, to ensure the lender is making a sound investment. Sometimes, the appraised value comes in lower than expected due to:

- **Market Shifts** – Prices may have risen too quickly, and comparable sales (recent home sales used to determine value) haven't caught up.
- **Limited Comparable Sales** – If there aren't enough recent sales nearby, the appraiser has limited data to support the purchase price.
- **Overpriced Offer** – Buyers sometimes offer more than a home is worth in the lender's eyes, especially in competitive markets.
- **Appraiser's Subjectivity** – Two different appraisers might value the same home differently based on their analysis.

Lenders will only loan based on the appraised value—not the agreed sale price—so a low appraisal creates a financing gap that needs to be addressed.

Step 2: What Are Your Options If the Appraisal Is Low?
A low appraisal doesn't have to end the deal. Here are five ways to move forward:

1. **Renegotiate with the Seller**
   - Ask the seller to lower the price to match the appraisal.
   - Many sellers will adjust rather than lose the deal, but this depends on market conditions.
2. **Challenge the Appraisal** (Rare, but possible)
   - If you believe the appraisal was inaccurate, your lender can request a review or a second opinion.
   - Provide better comparable sales or highlight property features the appraiser may have overlooked.
3. **Cover the Difference in Cash**
   - If you really want the home, you can pay the difference between the appraisal and purchase price.
   - This requires extra cash upfront but may be worth it in a competitive market.
4. **Split the Difference**
   - Negotiate a compromise where the seller lowers the price slightly, and the buyer covers part of the gap in cash.
   - Works well if both parties want to keep the deal moving forward.
5. **Walk Away (If You Have an Appraisal Contingency)**
   - If there's no agreement, you can back out of the deal if your contract includes an appraisal contingency—a clause that allows the buyer to cancel if the home appraises for less than the purchase price.
   - This protects your earnest money deposit—the good faith money you put down to secure the contract—and lets you look for another home.

Pro Tip: Some buyers include **appraisal gap coverage** in their offer, meaning they commit in advance to covering a portion of the gap. This can make their offer more competitive in a bidding war.

Final Thoughts:
A low appraisal doesn't have to kill your deal. Whether you renegotiate, challenge the appraisal, or bring extra cash to the table, the key is staying flexible and working with your real estate agent to find the best solution. If you're dealing with a low appraisal and need expert advice, reach out—I'm happy to help.

Preparing for Closing and Homeownership

VIDEO 20 – Credit Score Best Practices While Under Contract

Opening:

If you're under contract on a home, congratulations! But before you celebrate too much, let's talk about something that could impact your ability to close: your credit score. Many buyers don't realize that certain financial moves can jeopardize their mortgage approval. In this video, I'll cover best practices to keep your credit score stable—and the things you should avoid—while you're under contract.

Step 1: What You SHOULD Do
To keep your credit in good standing during this time:

- **Continue making all payments on time** – This includes rent, credit cards, car loans, and student loans.
- **Keep your credit card balances low** – Avoid maxing out any cards, even if you plan to pay them off.
- **Maintain your current job and income** – Lenders verify employment before closing, so stability is key.
- **Communicate with your lender** – If you have questions about financial decisions, check with them first.

Step 2: What You Should AVOID
Some actions might seem harmless but can actually derail your mortgage approval. Avoid:

- **Opening new lines of credit** – No new credit cards, car loans, or personal loans!
- **Making large purchases on credit** – This includes furniture, appliances, and vacations.
- **Co-signing for someone else's loan** – This affects your debt-to-income ratio, which lenders closely monitor.

- **Making big bank account transfers** – Large unexplained deposits or withdrawals could trigger red flags.
- **Closing old credit accounts** – This can reduce your credit history length and impact your score.

Step 3: Why This Matters
Your lender will re-check your credit before closing. If your score drops or your financial situation changes, it could:

- Delay your closing
- Result in a higher interest rate
- Even cause your loan approval to be revoked

Final Thoughts:
Buying a home is exciting, but keeping your financial profile stable is **critical** until you have those keys in hand. If you have any questions about what's safe to do—or what's not—reach out to me or your lender before making any changes. I'm here to help ensure a smooth closing and a successful home purchase!

VIDEO 21 – Life Preparation While Under Contract

Opening:
You're under contract on your new home—congratulations! While we work through the final steps toward closing, now is the perfect time to start preparing for your big move. In this video, I'll cover the essential things you should plan for before closing to ensure a smooth transition into your new home.

Step 1: Arranging Movers and Moving Logistics
Now is the time to start thinking about how you'll move:

- Research and book professional movers or a rental truck.
- Start packing non-essential items to reduce stress later.
- Notify your current landlord (if renting) about your move-out date.
- Arrange for help from friends or family if you're doing a DIY move.

Step 2: Setting Up Utilities and Services
Before moving in, you'll need to get your utilities set up:

- Contact your electric, gas, and water providers to transfer service.
- Arrange for internet, cable, and home security installations in advance.
- If your new home has HOA dues, set up automatic payments to avoid missing deadlines.

Step 3: Changing Your Address
Make sure to update your address with:

- The **U.S. Postal Service** to forward mail.

- Your **bank, credit cards, and insurance providers**.
- Your **employer and any subscriptions** that send physical mail.
- The **DMV and voter registration** if you're moving to a different county or state.

Step 4: Planning for Home Maintenance and Essentials
Owning a home comes with new responsibilities, so be prepared by:

- Creating a home maintenance checklist for things like HVAC filter changes and lawn care.
- Budgeting for necessary repairs or upgrades you want to make soon after moving in.
- Purchasing essential tools like a lawnmower, basic toolkit, and emergency supplies.

Step 5: Reviewing Your Closing Day Details
As we get closer to closing, you'll need to:

- Confirm your closing date, time, and location with your lender and title company.
- Review your final closing disclosure so you understand all costs.
- Make sure your funds for closing costs are ready and wired on time.

Final Thoughts:
Preparing for life in your new home starts **before** closing day. By taking these steps now, you'll have a much smoother transition. If you have any questions or need recommendations for movers, utility providers, or home services, I'm happy to help. Let's get you fully prepared for homeownership!

VIDEO 22 – Preparing for Closing: Settlement Statement

Opening:
You're almost at the finish line—closing day is just around the corner! But before you officially become a homeowner, there are a few important steps to take. In this video, I'll walk you through how to review your settlement statement, what to expect in the final days before closing, and how to ensure a smooth transition.

Step 1: Understanding the Settlement Statement
Your lender or title company will provide a **Closing Disclosure (CD)** or **settlement statement** a few days before closing. This document outlines:

- **The final purchase price**
- **Loan terms and interest rate**
- **Closing costs and lender fees**
- **Prorated taxes and insurance payments**

It's crucial to review this document carefully to make sure everything matches what you agreed to. If you see any discrepancies, let your lender or real estate agent know right away.

Step 2: Questions to Ask When Reviewing Your Settlement Statement
Before signing on the dotted line, ask:

- Are there any unexpected charges or fees?
- Does the final loan amount and interest rate match what I was promised?
- Have all seller concessions and agreed-upon credits been applied?
- What amount do I need to bring to closing, and how should I pay? (Wire transfer, cashier's check, etc.)

Step 3: Final Steps Before Closing
In the final days before closing, make sure to:

- **Conduct a final walkthrough** of the home to ensure everything is in the agreed-upon condition.
- **Confirm your closing date, time, and location** with your real estate agent and title company.
- **Arrange for home insurance** and provide proof to your lender.
- **Transfer utilities** into your name so you have power and water on move-in day.

Step 4: What to Expect on Closing Day
On closing day, you'll:

- Sign all final paperwork with your lender and title company.
- Pay any remaining closing costs or down payment.
- Receive the keys and celebrate—because you're now a homeowner!

Final Thoughts:
Closing is an exciting time, but it's important to be prepared. Reviewing your documents carefully, asking the right questions, and planning ahead will help ensure a stress-free closing. If you have any concerns or need guidance, reach out—I'm here to help every step of the way!

VIDEO 23 – Going to Closing: What to Expect and Best Practices

Opening:
The big day is finally here—**closing day!** This is when you officially become a homeowner. In this video, I'll explain what to expect at the closing table, key things to bring, and best practices to ensure everything goes smoothly.

Step 1: What Happens at Closing?
At closing, you'll sit down with your real estate agent, lender, and title company representative to sign all the final paperwork. This includes:

- **Loan documents** – Finalizing your mortgage terms.
- **Title transfer documents** – Making your ownership official.
- **Closing disclosure review** – Ensuring all numbers match what you expected.

Once everything is signed and your lender funds the loan, the keys to your new home will be yours!

Step 2: What to Bring to Closing
To avoid delays, make sure to bring:

- A **government-issued photo ID** (driver's license or passport).
- A **cashier's check or wire confirmation** for any remaining closing costs.
- Your **closing disclosure** to compare with final figures.
- Any **other required documents** your lender or title company requested.

Step 3: Questions to Ask at Closing
Before you sign, take a moment to ask:

- Does everything match what we agreed upon?
- When will I receive my keys?
- How do I make my first mortgage payment?
- What should I do if I have any issues after closing?

Step 4: Best Practices for a Smooth Closing
To ensure closing day goes as planned:

- **Review your documents beforehand** – You'll likely get them at least three days prior.
- **Arrive on time** and be prepared for the process to take an hour or more.
- **Double-check wire transfer instructions** with your title company to avoid fraud.
- **Celebrate**—you did it! You're officially a homeowner!

Final Thoughts:
Closing day is the last step before moving into your new home. By being prepared and asking the right questions, you can ensure everything goes smoothly. If you have any last-minute concerns, don't hesitate to reach out. Congratulations on your new home!

VIDEO 24 – After Closing – Now What?

Opening:

Congratulations, you're officially a homeowner! Now that closing is done and you have the keys, what's next? In this video, I'll talk about the essential steps to take after closing to make sure your transition into homeownership is smooth and stress-free.

Step 1: Secure Your New Home

One of the first things you should do is change the locks and security codes. You never know who might have copies of the previous keys. Consider installing a smart lock or a security system for added peace of mind.

Step 2: Set Up Utilities and Services

If you haven't already, make sure your electricity, water, gas, internet, and trash services are set up in your name and ready to go. You don't want to spend your first night in your new home without power or water!

Step 3: Organize Important Documents

Keep all of your closing documents, including your **warranty deed, mortgage papers, and title insurance**, in a safe place. If possible, create digital copies for easy access.

Step 4: Update Your Address

Now that you've moved in, don't forget to update your address with the postal service, your bank, credit card companies, and any subscriptions. Also, update your driver's license and voter registration if required.

Step 5: Plan for Home Maintenance

Regular maintenance will keep your home in great shape. Some key things to do right away include:

- Testing smoke and carbon monoxide detectors.
- Checking HVAC filters and scheduling routine servicing.
- Identifying the location of your water shut-off valve and electrical panel.

Step 6: Meet Your Neighbors and Explore Your Community

Building good relationships with your neighbors can make a big difference in feeling at home. Take some time to introduce yourself and explore local amenities, shops, and parks.

Final Thoughts:

Buying a home is a huge milestone, and taking these steps will help you start off strong. If you have any questions or need recommendations for home services, reach out—I'm always here to help. Enjoy your new home!

VIDEO 25 – Building Wealth Through Homeownership (Why Your First Home is Just the Beginning of Your Real Estate Journey)

Opening:

Congratulations, you're officially a homeowner! But buying a home isn't just about having a place to live – it's also one of the best ways to build long-term wealth. In this final video, we'll talk about how homeownership can set you up for financial success, future investments, and even early retirement.

Step 1: Understanding Equity and Why It's Powerful

Every mortgage payment you make helps build equity – the difference between your home's value and what you owe. Over time, as property values appreciate and you pay down your loan, you gain a financial asset that can be used for future investments, major expenses, or retirement.

Key Benefits of Building Equity:

- You can borrow against it for home renovations or emergencies.
- You can use it as a down payment on an investment property.
- The more equity you build, the more financial security you have.

Step 2: When and How to Move Up to Your Next Home

Your first home might not be your forever home, and that's okay! Many homeowners "move up" within 5-7 years to a larger home, a better neighborhood, or even an investment property.

Signs It Might Be Time to Upgrade:

- You need more space (new family members, working from home, lifestyle changes).
- You want to live in a different area with better schools, amenities, or commute times.
- Your home's value has significantly appreciated, and selling could help you buy a larger or better home.

Step 3: Should You Turn Your First Home into a Rental?

Instead of selling, some homeowners choose to rent out their first home and turn it into an investment property. This creates passive income and long-term wealth.

Pros of Keeping Your First Home as a Rental:

- It appreciates in value while tenants pay down your mortgage.
- It provides monthly rental income, which can supplement your finances.
- You diversify your assets and set yourself up for future investments.

How to Decide if This is Right for You:

- Does the rental income cover your mortgage and expenses?
- Are you prepared for property management responsibilities?

- Would selling now give you a better financial return?

Final Thoughts:
Your first home isn't just a place to live – it's a stepping stone to financial freedom and real estate opportunities. By being strategic, understanding equity, and planning for your next move, you can turn your home into an asset that works for you. If you're curious about what's next – whether it's selling, investing, or upgrading – let's talk! I'd love to help you plan your long-term real estate goals and make sure you get the most out of homeownership.

Chapter 5 Summary

Buying a home is one of the biggest financial decisions a person will make, and well-informed buyers feel more confident and prepared. This chapter explored how real estate agents can use AI-powered tools and a structured video library to educate clients efficiently, saving time while building trust.

The **Buyer Video Library** provides a 25-video series covering every step of the home-buying process, from onboarding and financing to negotiating contracts and closing. These videos serve as an essential tool for preparing buyers before consultations, guiding them through key milestones, and reducing repetitive questions.

Key topics include getting pre-qualified for a mortgage, understanding closing costs, making competitive offers, handling multiple-offer situations, and navigating inspections, appraisals, and contingencies. Buyers also learn how to review HOA documents, title policies, and homeowners' insurance to avoid costly surprises.

This chapter also highlighted the importance of preparing for closing and post-closing responsibilities, such as securing utilities, organizing documents, and maintaining the home.

By integrating AI-driven education and automation, real estate agents can scale their business while offering a high-value, client-centric experience. This approach empowers buyers, builds credibility, and strengthens long-term relationships—key components of success in real estate.

Author Note: While we are newer to AI ourselves, we created this entire chapter, from scratch, in just a few hours. You can get all the scripts online and modify them as you see fit! Use our videos or record your own.

6. SELLER PROCESS

Introduction: AI and Seller Education

Selling a home is one of the biggest financial transactions a homeowner will make, and a well-informed seller is far more likely to achieve a smooth, successful sale. As a real estate agent, your role isn't just to list homes—it's to educate, guide, and empower sellers to make the best decisions for their situation. However, answering the same questions repeatedly, managing multiple listings, and ensuring each client is fully prepared can be time-consuming. That's where AI-powered tools come in.

AI isn't just about automation—it's about enhancing the seller experience by providing accurate data, targeted marketing, and seamless communication. Instead of spending hours manually researching market conditions, AI can instantly generate pricing recommendations based on real-time market data. AI-powered CRMs can automate follow-ups, ensuring sellers are kept informed every step of the way. AI-driven marketing tools can also create high-quality listing descriptions, optimize ad placements, and track engagement to ensure the right buyers see the home.

This chapter is designed to show you how AI can improve every stage of the home-selling process—from pricing and listing preparation to negotiations and closing. You'll find a structured Seller Video Library, similar to the Buyer Video Library, that will help you educate clients at scale. By leveraging AI to provide better data, stronger marketing, and a streamlined experience, you can help sellers sell faster, for more money, and with less stress—while also making your business more efficient.

Seller Video Library

One of the biggest challenges for sellers is understanding the complexities of the home-selling process. From pricing their home correctly to preparing for showings, handling negotiations, and making it to closing, there are countless details to manage. As a real estate agent, providing clear, structured education upfront can save time, build trust, and lead to smoother transactions. That's where an AI-powered Seller Video Library comes in.

This video series is designed to educate sellers before they ever sit down with you, making your consultations more efficient and helping them feel more confident about the process. AI can assist in generating scripts, automating video creation, and even personalizing educational content based on client needs. You can use the QR codes to go to our website, where we have the completed video with a voice over all set to go. You'll also find the scripts that we used. You can copy the script and personalize it for your brand, and/or make specific changes as needed for your local market. By sending sellers a curated set of videos at different stages—before listing, during showings, and leading up to closing—you ensure they stay informed, reducing last-minute surprises and unnecessary delays.

Here's how this video series will help:

- **Save Time** – Instead of answering the same questions over and over, send a video and use your meetings for higher-value conversations.
- **Build Trust** – An educated seller is more likely to trust your recommendations on pricing, staging, and negotiation strategies.
- **Increase efficiency** – With pre-recorded AI-assisted video content, you can scale your education efforts while maintaining a personal touch.

In this chapter, we'll walk through a structured step-by-step seller process, with each stage supported by video content that can be easily shared with clients. By leveraging AI and video education, you can create a seamless experience for your sellers, setting yourself apart as a trusted expert in the market.

Seller Onboarding and Preparation

VIDEO 1 – Understanding the Home Selling Process

Opening

Thinking about selling your home? The process may seem overwhelming, but with the right approach and a trusted real estate agent, you can sell your home efficiently and for top dollar. In this video, we'll walk you through the key steps of the home-selling process so you know exactly what to expect from start to finish."

What to Expect

Selling a home involves multiple steps, from preparing your property to marketing it effectively, negotiating with buyers, and finally, closing the deal. The key is to stay organized, work with a professional, and use smart strategies—including AI-driven tools—to streamline the process.

Step 1: Hire the Right Real Estate Agent

Working with an experienced agent ensures that you get expert guidance, strong marketing, and skilled negotiation to get the best price for your home.

Step 2: Prepare Your Home for the Market

Decluttering, deep cleaning, and staging can make your home more appealing to buyers. A well-presented home sells faster and for more money.

Step 3: Price Your Home Correctly

Your agent will conduct a Comparative Market Analysis (CMA) to determine the best listing price based on recent sales, market conditions, and buyer demand.

Step 4: Market Your Home Effectively
High-quality photography, virtual tours, and social media exposure are essential. Your real estate agent's AI-powered marketing tools can ensure your listing reaches the right audience.

Step 5: Showings and Buyer Feedback
Allow flexible showing times, keep your home clean, and be prepared for buyer feedback. Your real estate agent's AI-driven scheduling tools can help manage this efficiently.

Step 6: Receive and Negotiate Offers
Your real estate agent will present and analyze all offers, helping you negotiate the best deal, considering price, contingencies, and closing terms.

Step 7: Home Inspection and Appraisal
Buyers will typically schedule an inspection and appraisal. Addressing issues proactively can help keep the deal moving forward smoothly.

Step 8: Closing the Deal
Final paperwork is signed, the home is transferred to the buyer, and you receive your proceeds. Congratulations—you've successfully sold your home!

Final Thoughts
Selling your home doesn't have to be stressful when you have a clear roadmap and a professional team by your side. If you're ready to start the process, reach out to your real estate agent today and let's get your home sold!

VIDEO 2 – How to Choose the Right Listing Agent

Opening
Choosing the right listing agent is one of the most important decisions you'll make when selling your home. A great agent can mean the difference between a quick, profitable sale and a long, frustrating process. In this video, we'll go over key factors to consider when selecting the best real estate agent for your needs.

What to Expect
Not all real estate agents are the same, and selecting the wrong one can cost you time and money. Your agent should be knowledgeable, proactive, and experienced in your local market. Here's how to find the right fit.

Step 1: Look for Experience and a Proven Track Record
Ask potential agents how many homes they've sold in your area, what their average days on market are, and if they specialize in your property type.

Step 2: Evaluate Their Marketing Strategy
Your agent should have a solid marketing plan, including high-quality listing photos, virtual tours, social media marketing, and AI-driven advertising to maximize exposure.

Step 3: Check Reviews and References
Look up online reviews and ask for past client testimonials. A great agent should have a history of satisfied clients.

Step 4: Communication and Availability
Your agent should be responsive, available when needed, and provide clear, honest communication throughout the process.

Step 5: Understand Their Commission and Contract Terms
Ask about commission rates and any contract requirements before signing. The best agent will be transparent about costs and expectations.

Final Thoughts
The right real estate agent will be your partner throughout the entire home-selling process. Take the time to interview multiple agents, ask the right questions, and choose someone you trust to get the job done right.

VIDEO 3 – Pre-Listing Home Inspection: Is It Worth It?

Opening:
Most sellers expect a home inspection when a buyer makes an offer—but what if you got one before listing your home? A pre-listing inspection can help you avoid surprises, strengthen your position in negotiations, and even speed up your sale. Let's discuss why this can be a smart move.

What to Expect:
A pre-listing inspection is a voluntary home inspection done before you list your home. It allows you to identify and fix potential issues before buyers see them, making your home more attractive and reducing the chances of last-minute deal-breakers.

Step 1: Hire a Licensed Home Inspector.
Choose a reputable professional who will provide a thorough report.

Step 2: Review the Inspection Report.
Identify any red flags that could delay or jeopardize your sale.

Step 3: Decide What to Fix.
Address major concerns like roofing, plumbing, or electrical issues to boost buyer confidence.

Step 4: Use the Report to Your Advantage
Show buyers that your home is in great condition or disclose minor issues upfront to avoid renegotiations later.

Final Thoughts:
A pre-listing inspection isn't required, but it **can make your home stand out in a competitive market.** If you want fewer surprises and stronger offers, this is an investment worth considering!

VIDEO 4 – The Top Home Repairs to Maximize Value

Opening:
Before you list your home, it's important to make smart repairs that will boost its value and attract buyers. But not all upgrades are worth the money—some offer a much higher return on investment than others. Let's go over the top repairs that make the biggest impact.

What to Expect:
Buyers love move-in-ready homes. Fixing obvious issues upfront helps your home sell faster and for a higher price. We'll focus on repairs that add value without unnecessary spending.

Top Repairs to Maximize Value:

1. **Paint & Touch-Ups** – A fresh coat of neutral paint makes your home look newer and more inviting.
2. **Kitchen Updates** – Replacing cabinet hardware, light fixtures, or countertops can modernize your space without a full remodel.
3. **Bathroom Improvements** – Fix leaky faucets, recaulk tubs, and update mirrors for an easy refresh.
4. **Flooring Fixes** – Replace old carpets, polish hardwood floors, and fix damaged tiles.
5. **Curb Appeal Enhancements** – A well-maintained lawn, trimmed bushes, and a clean entryway make a great first impression.
6. **Update Lighting** – Replace dated and dim light fixtures is an inexpensive way to brighten up a space.

Final Thoughts:
Small, strategic updates can make a big difference in how quickly your home sells and how much buyers are willing to pay. Focus on repairs that matter most, and let's get your home ready for the market!

VIDEO 5 – Staging Your Home for Maximum Buyer Appeal

Opening:
Did you know that staged homes sell faster and for more money than vacant or cluttered homes? Staging helps buyers visualize themselves living in your space, making your home more appealing. Whether you hire a professional or stage your home yourself, let's go over what works best.

What to Expect:
Staging isn't about decorating—it's about highlighting your home's best features and making it feel welcoming to buyers. The goal is to create a clean, neutral, and inviting atmosphere that appeals to the widest audience.

Step 1: Declutter & Depersonalize – Remove excess furniture, family photos, and personal items to create a clean slate.

Step 2: Optimize Furniture Placement – Arrange furniture to make rooms feel spacious and functional.

Step 3: Enhance Lighting and Ambiance – Use natural light, add lamps, and create a warm, inviting feel.

Step 4: Add Small Touches – Fresh flowers, stylish throw pillows, and tasteful décor can make a big impact.

Step 5: Boost Curb Appeal – A welcoming exterior sets the tone for the entire showing experience.

Final Thoughts:
A well-staged home looks better in photos, attracts more buyers, and sells for more money. Even small adjustments can make a big difference. Take the time to stage your home properly, and let's make that first impression count!

Pricing and Market Strategy

VIDEO 6 – How to Price Your Home Right the First Time

Opening:
Pricing your home correctly from the start is one of the most important factors in getting it sold quickly and for top dollar. **Overpricing can scare away buyers, while underpricing can leave money on the table.** In this video, we'll break down the pricing strategy that gets you the best results.

What to Expect:

We'll cover the dangers of overpricing, how market conditions influence pricing, and how AI-powered tools help determine the best price for your home. The goal is to attract serious buyers while maximizing your profit.

Step 1: Understand the Market Conditions – Is it a buyer's or seller's market? This impacts how aggressive or conservative your pricing should be.

Step 2: Use a Comparative Market Analysis (CMA) – Looking at similar home sales in your area helps determine a competitive price.

Step 3: Consider Buyer Perception – Buyers compare homes side by side, so your price must align with what they expect for the area.

Step 4: Leverage AI Pricing Tools – AI can analyze real-time data, trends, and even buyer behavior to suggest an optimal price.

Step 5: Set a Price That Attracts Offers – Pricing slightly below market value in a competitive market can spark multiple offers and increase your final sale price.

Final Thoughts:
Pricing isn't about guessing—it's about data, strategy, and market demand. Work with a knowledgeable agent, use AI-driven pricing tools, and set a price that brings in serious buyers right from the start.

VIDEO 7 – The Power of Comparative Market Analysis (CMA)

Opening:
How do you know what your home is worth? A Comparative Market Analysis (CMA) is one of the most effective ways to determine a competitive price. A CMA compares your home to similar properties that have recently sold, are currently on the market, or didn't sell—giving you a realistic pricing range.

What to Expect:
We'll walk through how a CMA works, what factors influence your home's value, and how buyers use CMAs to compare listings.

Understanding a CMA:
Step 1: Look at Recent Sales (Sold Comps) – These show what buyers are actually willing to pay in today's market.

Step 2: Analyze Active Listings (Your Competition) – These are the homes your buyers are also considering. Your home must be competitively priced against them.

Step 3: Check Expired and Withdrawn Listings – If homes didn't sell, pricing or condition was likely an issue. Avoid these mistakes.

Step 4: Adjust for Unique Features – Your home's size, upgrades, location, and condition all affect how it compares to others.

Step 5: Use AI to Refine Pricing – AI can analyze thousands of data points, making CMA reports even more accurate and tailored to market trends.

Final Thoughts:
A well-prepared CMA helps you price your home confidently and avoid costly pricing mistakes. Work with your real estate agent to analyze this data and use AI tools to get the most precise estimate possible.

VIDEO 8 – Common Pricing Mistakes and How to Avoid Them

Opening:
Did you know that **overpricing is the #1 reason homes sit on the market too long**? Many sellers make critical pricing mistakes that turn buyers away, lead to price reductions, and ultimately cost them money. In this video, we'll go over the biggest pricing mistakes and how to avoid them.

What to Expect:
We'll cover why "testing the market" with a high price backfires, why price reductions can hurt your sale, and how buyers perceive pricing.

Top Pricing Mistakes to Avoid:

1. **Overpricing from the Start** – Buyers compare prices instantly online. If your home is too high, they won't even look at it.
2. **"Testing the Market" with a High Price** – Homes that start overpriced sit too long and often sell for less than market value in the end.
3. **Making Incremental Price Drops** – Multiple small reductions signal **desperation**, making buyers wait for an even lower price.
4. **Ignoring Market Trends** – If the market is shifting, your pricing strategy must adapt to buyer demand.
5. **Pricing Based on Emotion Instead of Data** – Your home is personal, but buyers only see market value. A CMA and AI-driven pricing tools help keep pricing objective.

Final Thoughts:
The right price attracts buyers. **The wrong price keeps them away.** Avoid these common mistakes, rely on market data and AI-driven pricing, and set a price that gets your home sold fast and for top dollar.

VIDEO 9 – How AI Helps Predict Market Trends for Sellers

Opening:
Imagine if you could predict buyer demand before listing your home. AI-powered tools now allow sellers to analyze market conditions with incredible accuracy, helping you price your home strategically, market it effectively, and sell faster.

What to Expect:
We'll explore how AI predicts buyer demand, pricing trends, and market shifts to give you a competitive edge when selling your home.

How AI Helps Predict Market Trends:

1. **Analyzes Local and National Data** – AI compares past and present home sales, mortgage rates, and inventory levels to detect trends.
2. **Predicts Buyer Behavior** – AI tools study search patterns, online engagement, and buyer activity to forecast which homes will sell fastest.
3. **Optimizes Pricing Strategies** – AI-powered pricing models adjust in real-time based on market demand and competing listings.
4. **Identifies the Best Time to Sell** – AI helps pinpoint seasonal trends and high-demand periods, so you list at the right time.
5. **Enhances Marketing Strategies** – AI determines which listing features buyers engage with most, so you know how to highlight your home's best-selling points.

Final Thoughts:
AI isn't just a tech trend—it's changing the way homes are sold. By using AI-driven market analysis, you stay ahead of the competition and make smarter pricing and marketing decisions.

Marketing and Listing Exposure

VIDEO 10 – How Professional Photography and Videos Help Sell Faster

Opening:
In today's digital world, first impressions happen online. Buyers are scrolling through listings, and if your photos and videos don't stand out, they'll move on to the next home. Homes with professional photography sell faster and for more money, and in this video, we'll explain why investing in great visuals is essential.

What to Expect:

We'll cover the role of high-quality photos, video tours, and virtual staging in attracting more buyers and creating a competitive advantage.

To Make Your Listing Visually Stand Out, Your Agent Can:

1. **Hire a Professional Photographer** – Professional photos are bright, clear, and showcase your home in the best possible light.
2. **Include a Video Tour** – Video walkthroughs give buyers a real feel for the home, increasing engagement and interest.
3. **Consider Virtual Staging** – AI-powered virtual staging can help furnish an empty home, making it more appealing online.
4. **Use Drone Photography for Exterior Shots** – Aerial images showcase your home's location, neighborhood, and lot size.
5. **Optimize Your Listing Photos** – Arrange photos strategically to highlight the best features first, drawing buyers in.

Final Thoughts:
Online home searches are the first showing, so it's important to make sure a listing grabs buyers' attention instantly. Investing in professional visuals is one of the best ways to do just that!

VIDEO 11 – Creating an Engaging Online Listing That Stands Out

Opening:
Your online listing description is just as important as your photos—it's what brings your home to life for potential buyers. A compelling listing highlights key features, tells a story, and makes buyers feel emotionally connected before they even schedule a showing.

What to Expect:
We'll cover how a powerful listing description can grab attention and get buyers excited about your home.

Your Real Estate Agent Crafts a Winning Listing Description:

1. **Starts with a Strong Opening Line** – Instead of "Charming 3-bedroom home," try "Stunning, sun-filled home with breathtaking views!"
2. **Highlights Unique Features** – Mentions high-end appliances, hardwood floors, smart home features, or a finished basement.
3. **Paints a Picture with Descriptive Language** – Instead of "large backyard," say "a spacious, fully fenced backyard perfect for summer gatherings."
4. **Uses AI to Optimize Wording** – AI tools can refine descriptions and suggest keywords that buyers are searching for.
5. **Includes a Call to Action** – Encourage buyers to schedule a tour today before this home is gone!

Final Thoughts:
Your listing should create an emotional connection and make buyers want to see more. A well-crafted description, paired with stunning visuals, can set your home apart in a crowded market.

VIDEO 12 – The Role of Social Media in Selling Your Home

Opening:
Social media isn't just for staying connected—it's a powerful marketing tool that can get your home in front of thousands of potential buyers. Facebook, Instagram, and even TikTok are now key platforms where buyers discover homes for sale. If your home isn't on social media, you're missing out on a huge opportunity!

What to Expect:
We'll cover how social media marketing expands your home's reach, how AI-powered ads target the right buyers, and how video content makes listings more engaging.

How Your Real Estate Agent Can Maximize Your Home's Social Media Exposure:

1. **Posts Stunning Photos and Videos** – High-quality visuals get more likes, shares, and engagement.
2. **Uses AI-Powered Targeting for Ads** – AI tools help real estate agents place ads in front of buyers actively searching for homes like yours.
3. **Creates Short Video Walkthroughs** – Videos on Instagram Reels, TikTok, and Facebook Stories get more engagement than static images.
4. **Encourages Shares and Engagement** – Ask friends, family, and followers to share your post to increase visibility.
5. **Uses AI-Generated Captions & Hashtags** – AI can help craft engaging descriptions and trending hashtags to boost your post's reach.

Final Thoughts:
Social media marketing isn't optional—it's essential. The more people see your home online, the more interest and offers you'll receive. Leverage AI-powered marketing strategies to get your home sold faster!

VIDEO 13 – Understanding MLS and Syndication for Exposure

Opening:
When you list your home, it doesn't just appear on one website—it gets syndicated across dozens of real estate platforms like Zillow, Realtor.com, and Redfin. This massive exposure increases your chances of finding the right buyer quickly and at the best price.

What to Expect:
We'll explain how MLS (Multiple Listing Service) works, how syndication expands your reach, and why online exposure is key to a fast sale.

Step 1: List on the MLS – The MLS is the main database where real estate agents list properties, ensuring it reaches other agents and buyers.

Step 2: Ensure Your Listing is Syndicated – When listed properly, your home will appear on major real estate sites automatically.

Step 3: Monitor Online Listing Performance – Your real estate agent can use AI tools to track listing views, buyer inquiries, and engagement levels.

Step 4: Update Your Listing Regularly – Making changes? Your agent can incorporate fresh updates, price changes, or new photos to boost visibility on listing platforms.

Step 5: Leverage AI-Driven Listing Optimization – Your agent can utilize AI to analyze which photos perform best and how to improve your listing's ranking.

Final Thoughts:
The MLS and listing syndication make sure your home reaches the right buyers quickly. Pairing this with AI-driven listing strategies can increase your home's visibility and attract serious offers!

Showings and Buyer Interaction

VIDEO 14 – How to Prepare for Showings and Open Houses

Opening:
When buyers walk through your home, **first impressions are everything**. A well-prepared home makes buyers feel welcome, increases their emotional connection, and leads to stronger offers. In this video, we'll go over a step-by-step checklist to ensure your home is show-ready every time.

What to Expect:
We'll cover what to clean, what to hide, how to stage last-minute, and how to create a welcoming atmosphere for buyers.

Step 1: Declutter and Depersonalize – Remove personal items, clear countertops, and minimize decor to make the space feel neutral.

Step 2: Deep Clean Everything – Floors, windows, bathrooms, and kitchens should sparkle. **A clean home feels bigger and more inviting.**

Step 3: Let in Natural Light – Open blinds and curtains to create a bright, airy atmosphere that buyers love.

Step 4: Set the Right Temperature and Smell – Keep the home comfortable and fresh-smelling—no overpowering air fresheners!

Step 5: Secure Valuables & Personal Items – Store away any valuable items, personal documents, and prescription medications.

Step 6: Make Last-Minute Touches – Add fresh flowers, set the dining table, or put out a bowl of lemons for a subtle, homey touch.

Step 7: Leave the House – Buyers need space to explore and envision themselves in your home. **Always leave during showings.**

Final Thoughts:
A show-ready home makes a great impression and helps buyers feel excited about making an offer. Keep your home clean, bright, and neutral, and you'll set yourself up for a faster, more profitable sale!

VIDEO 15 – What Buyer Feedback Tells You and How to Adjust

Opening:
When buyers tour your home, their feedback is one of your most valuable tools. Are buyers excited, or are they hesitant? Are they saying the home is priced too high, or is something turning them off? Understanding and acting on buyer feedback can help you sell faster.

What to Expect:
We'll explain how to interpret buyer feedback, what common objections mean, and how to make adjustments that attract stronger offers.

Analyzing and Acting on Buyer Feedback:
Step 1: Look for Patterns – If multiple buyers mention a problem (like a dark room or outdated kitchen), it's worth addressing.

Step 2: Pay Attention to Pricing Feedback – If buyers are interested but not making offers, **your price may be too high.**

Step 3: Listen to Emotional Reactions – Buyers should feel excited and connected—if they don't, staging or minor updates might help.

Step 4: Improve Based on Specific Comments – Example: If buyers say a room feels too dark, add better lighting or lighter curtains.

Step 5: Use AI to Track & Analyze Feedback – AI-powered CRMs can aggregate feedback and help spot trends more quickly.

Final Thoughts:
Buyer feedback isn't criticism, it's insight! Small adjustments based on feedback can turn hesitant buyers into serious offers. Work with your agent to track feedback and make smart, strategic updates.

VIDEO 16 – How AI Helps Optimize Showing Schedules

Opening:
Coordinating showings and following up with buyers may be time-consuming, but AI can make the process easier, faster, and more effective. With AI-driven scheduling tools and automated follow-ups, your agent can ensure every interested buyer gets the attention they need.

What to Expect:
We'll discuss how AI streamlines showing coordination, tracks buyer interest, and automates follow-ups to keep buyers engaged.

How AI Improves Showings & Follow-Ups:

1. **AI-Powered Showing Scheduling** – AI tools help coordinate showings with buyers, agents, and homeowners to reduce conflicts.
2. **Smart Lead Tracking** – AI can rank buyers based on interest and engagement, helping your agent focus on the most serious leads.
3. **Automated Follow-Ups** – AI-powered CRMs send personalized texts and emails to keep buyers interested after a showing.
4. **Predictive Buyer Insights** – AI can analyze buyer behavior and feedback to suggest which leads are most likely to make an offer.
5. **Appointment Reminders and Updates** – AI sends reminders to buyers about scheduled showings, ensuring maximum attendance.

Final Thoughts:
AI can make showing coordination effortless and ensures no potential buyer slips through the cracks. By automating the scheduling and follow-up process, your real estate agent will increase buyer engagement and boost your chances of receiving strong offers.

Offers and Negotiation

VIDEO 17 – Understanding the Offer Process for Sellers

Opening:
Receiving an offer on your home is exciting—but not all offers are created equal. Beyond the price, there are key contract terms that can make or break a deal. In this video, we'll discuss how to evaluate offers, understand contingencies, and choose the best one for your situation.

What to Expect:
We'll break down how to read an offer, what contingencies mean, and how to compare multiple offers beyond just the price.

Step 1: Review the Offer Price, But Look Deeper – The highest price isn't always the best offer if the terms are weak.

Step 2: Understand Contingencies – Common contingencies include financing, appraisal, inspection, and selling the buyer's current home.

Step 3: Check the Buyer's Financing Strength – A pre-approved buyer with a strong loan is more reliable than one still seeking financing.

Step 4: Look at Earnest Money and Closing Timeline – A higher earnest money deposit shows serious commitment. The closing date should align with your needs.

Step 5: Consider Concessions & Repairs – Does the buyer request seller-paid closing costs or pre-closing repairs? Factor these into your decision.

Final Thoughts:
Choosing the right offer means balancing price, terms, and risk. Your agent will help guide you through the process to ensure you accept the offer that best protects your interests and gets you to the closing table smoothly.

VIDEO 18 – How to Negotiate the Best Deal for Your Home

Opening:
The first offer isn't always the final offer! Negotiation is key to maximizing your sale price and securing the best terms. Whether handling counteroffers, lowball offers, or multiple bids, having a strong negotiation strategy can make all the difference.

What to Expect:

We'll go over common negotiation tactics, how to respond to different types of offers, and how to use leverage to your advantage.

Negotiating the Best Deal:

1. **Stay Calm and Professional** – Negotiations aren't personal; they're part of the process. **Keep emotions out of it.**
2. **Respond Strategically to Lowball Offers** – Instead of rejecting outright, counter with a reasonable price and strong terms.
3. **Leverage Multiple Offers When Possible** – If you have multiple offers, use them to create a bidding war.
4. **Negotiate More Than Just Price** – Closing costs, move-out dates, and contingencies can all be adjusted in your favor.
5. **Use AI-Driven Market Data to Justify Your Price** – AI-powered tools can provide real-time market trends that back up your counteroffer.

Final Thoughts:
Successful negotiation isn't just about price—it's about securing the best overall deal. Your agent's expertise and market insights will help you navigate negotiations and walk away with the strongest possible contract.

VIDEO 19 – What Happens During a Home Inspection

Opening:
Most buyers will request a home inspection, and it can be a make-or-break moment in the sale process. Knowing what to expect and how to prepare can help you avoid costly surprises and keep the deal moving forward.

What to Expect:
We'll cover what happens during the inspection, common buyer concerns, and how to handle repair requests effectively.

Step 1: Make Small Repairs Before the Inspection – Fix leaky faucets, missing caulk, loose door handles, and other minor issues in advance.

Step 2: Ensure Easy Access to Key Areas – The inspector needs access to attics, basements, HVAC systems, and electrical panels.

Step 3: Understand Common Inspection Findings – Issues like roof wear, plumbing leaks, and HVAC concerns may come up.

Step 4: Be Ready for Buyer Requests – Buyers may ask for repairs, closing cost credits, or price reductions based on the report.

Step 5: Use AI to Assess Repair Costs and Prioritize Fixes – AI-powered cost estimators can help you decide what's worth fixing versus negotiating.

Final Thoughts:
A successful inspection is all about preparation. Handling potential issues before they arise can prevent delays and keep your sale on track.

VIDEO 20 – The Appraisal Process and What It Means for Your Sale

Opening:
Even if you agree on a price with a buyer, the appraisal must support that price, or the deal could be in jeopardy. Understanding how the appraisal process works can help you avoid surprises and ensure your sale goes smoothly.

What to Expect:
We'll explain how appraisals impact home sales, potential challenges, and what happens if your home under-appraises.

Step 1: Understand How Appraisers Determine Value – Appraisers compare recent home sales, market trends, and your home's condition.

Step 2: Prepare Your Home for the Appraisal – A clean, well-maintained home with documented upgrades can positively influence the report.

Step 3: Provide Comps and Upgrades to the Appraiser – Your agent can share comparable sales and a list of improvements to justify your price.

Step 4: Know Your Options If the Appraisal is Low – You can negotiate with the buyer, challenge the appraisal, or adjust the sale price.

Step 5: Use AI-Powered Market Data to Support Pricing – AI-driven valuation tools can provide backup data to justify your home's value.

Final Thoughts:
The appraisal process doesn't have to derail your sale. By pricing correctly, preparing in advance, and knowing your options, you can reduce risk and keep your home sale on track.

Closing and Moving Forward

VIDEO 21 – What to Expect After Going Under Contract

Opening:

Congratulations! You're officially under contract—but the home sale isn't final just yet. There are still key steps and deadlines that need to be met before you reach the closing table. In this video, we'll explain what to expect after accepting an offer and how to ensure a smooth process.

What to Expect:
We'll cover the timeline of events, key responsibilities, and common issues that may arise during the contract period.

Step 1: Understand the Contract Timeline – Your contract will include specific deadlines for inspections, appraisals, and financing approval.

Step 2: Prepare for the Home Inspection – Expect the buyer to hire an inspector to evaluate your home's condition (see Video 19 for more details).

Step 3: Appraisal Process – If the buyer is financing the home, their lender will order an appraisal to confirm the home's value (see Video 20).

Step 4: Respond to Buyer Requests and Negotiations – The buyer may request repairs, credits, or price adjustments based on their findings.

Step 5: Work with the Title Company – The title company will verify ownership, clear any liens, and prepare closing documents.

Final Thoughts:
Once you go under contract, staying organized and meeting deadlines is key to keeping your sale on track. Work closely with your agent to navigate inspections, appraisals, and final negotiations so you can reach a successful closing.

VIDEO 22 – Navigating the Closing Process Like a Pro

Opening:
The closing process is the final step in selling your home. But what exactly happens at closing? What documents do you need to sign? In this video, we'll walk you through the closing process so you know exactly what to expect.

What to Expect:
We'll cover what happens at closing, what documents you'll need to sign, and how the funds are transferred.

Step 1: Review the Closing Disclosure – This document outlines your final numbers, including the sale price, fees, and net proceeds.

Step 2: Sign the Required Documents – You'll sign the deed transfer, settlement statement, and other required paperwork at closing.

Step 3: Pay Any Seller Fees – Closing costs for sellers may include agent commissions, title fees, and any agreed-upon concessions.

Step 4: Confirm Loan Payoff (If Applicable) – If you have an outstanding mortgage, your lender will receive the payoff from the sale proceeds.

Receive Your Sale Proceeds – Once everything is finalized, your funds will be transferred via wire or check—congratulations, your home is officially sold!

Final Thoughts:
The closing process should be straightforward if you've prepared ahead of time. Work closely with your real estate agent, review documents in advance, and be ready to sign on the dotted line for a smooth closing experience.

VIDEO 23 – Common Closing Delays and How to Prevent Them

Opening:

Not all closings go as planned. Delays can happen, but the good news is that most issues are avoidable with the right preparation. In this video, we'll go over the top reasons closings get delayed and how to keep your sale on track.

What to Expect:
We'll discuss common causes of closing delays and what steps you can take to prevent them.

Common Closing Delays and How to Avoid Them:

1. **Financing Issues** – If the buyer's loan gets delayed, stay in close contact with their lender to track progress.
2. **Low Appraisal** – If the appraisal comes in low, you may need to renegotiate the price or challenge the appraisal.
3. **Inspection-Related Delays** – Address any major repair issues quickly and efficiently to keep negotiations moving.
4. **Title Problems** – Liens or ownership disputes can delay closing, so work with the title company early to clear any issues.
5. **Last-Minute Buyer Issues** – Some buyers get cold feet or face unexpected life changes—stay flexible and have a backup plan.

Final Thoughts:
While delays can be frustrating, proper preparation and clear communication can help avoid most issues. Stay proactive, address concerns early, and work with your team to keep your sale on schedule.

VIDEO 24 – Moving Out: Timeline, Tips, and Final Responsibilities

Opening:
Now that you've sold your home, it's time to prepare for moving out! Whether you're relocating across town or across the country, having a clear plan will make the process much smoother.

What to Expect:
We'll go over key moving timelines, final walkthrough expectations, and how to hand over the keys.

Step 1: Confirm Your Move-Out Date – Work with the buyer and your agent to finalize when you'll be fully moved out.

Step 2: Schedule Movers Early – If hiring movers, book them in advance to avoid last-minute stress.

Step 3: Deep Clean and Repair Minor Issues – Leave the home in good condition, removing all belongings and repairing small damages.

Step 4: Buyer's Final Walkthrough – The buyer will do a walkthrough to ensure everything is as agreed upon in the contract.

Step 5: Hand Over Keys, Garage Openers and Documents – Leave all keys, remotes, and security system instructions for the new owner.

Final Thoughts:
A well-planned move reduces stress and ensures a smooth transition for both you and the buyer. Plan ahead, stay organized, and enjoy your next adventure!

VIDEO 25 – Life After Selling: What's Next?

Opening:
Selling your home is a big milestone, but what comes next? Whether you're buying another home, renting, or relocating, there are a few important post-sale steps you need to consider.

What to Expect:
We'll go over what happens after selling your home, including tax considerations, reinvesting profits, and next steps.

Step 1: Understand Potential Tax Implications – If you made a large profit, check with a tax professional about capital gains taxes.

Step 2: Reinvest or Save Your Proceeds – Consider investing in another home, rental property, or other financial opportunities.

Step 3: Update Your Address and Utilities – Notify banks, credit cards, and subscriptions of your new address.

Step 4: Store Important Closing Documents – Keep copies of your **final closing statement, deed, and tax forms.**

Step 5: Plan for Your Next Move – Whether buying or renting, have a clear financial and housing strategy moving forward.

Final Thoughts:
Selling your home isn't just about the sale—it's about what comes next. Take time to plan your next steps wisely so you can move forward with confidence and financial security.

Chapter 6 Summary

Selling a home is a complex process, and well-informed sellers make better decisions, leading to smoother transactions and higher sale prices. This chapter explores how real estate agents can leverage AI-powered tools and a structured video library to educate sellers efficiently, streamline the selling process, and build trust—saving time while enhancing client experience.

The Seller Video Library provides a 25-video series covering every stage of the home-selling journey, from onboarding and home preparation to pricing strategies, marketing, negotiations, and closing. These videos help sellers understand key milestones before consultations, reducing repetitive questions and ensuring they are better prepared to make informed choices.

Key topics include pricing a home correctly, using AI-driven market analysis, staging for maximum appeal, leveraging professional photography and social media marketing, and handling buyer negotiations. Sellers also gain insight into the contract-to-closing process, including home inspections, appraisals, common closing delays, and final move-out responsibilities.

This chapter emphasizes the importance of effective communication, data-driven decision-making, and AI-powered automation to optimize every step of the seller process. By integrating these tools, real estate agents can provide a high-touch, high-efficiency service, ensuring sellers feel supported while improving transaction speed and success rates.

By educating sellers proactively and utilizing AI to enhance pricing, marketing, and negotiation strategies, real estate agents differentiate themselves from competitors, increase client satisfaction, and build long-term relationships—key drivers of success in today's competitive real estate market.

Author Note: While we are newer to AI ourselves, we created this entire chapter, from scratch, in just a few hours. You can get all of the scripts online and modify them as you see fit! Use our videos or record your own.

7. ROLE PLAY WITH AI

In real estate, success isn't just about knowing the market—it's about how you communicate, negotiate, and build trust with clients. Whether you're a new agent trying to gain confidence or an experienced real estate agent looking to refine your skills, practicing conversations is key. Roleplaying is one of the most effective ways to improve, but many agents struggle to find consistent practice partners. This is where AI steps in as your 24/7, no-judgment coach to help you sharpen your skills and gain a competitive edge.

Even top professionals never stop practicing. Tiger Woods, one of the greatest golfers of all time, still works with a coach to fine-tune his swing. Likewise, elite salespeople, public speakers, and negotiators constantly refine their approach. The best real estate agents don't just rely on natural talent—they train, adjust, and perfect their conversations to handle objections, close deals, and serve clients better. AI-powered roleplaying allows agents to rehearse scripts, practice responses, and refine their delivery in a risk-free environment before speaking to real clients.

AI is revolutionizing roleplaying by providing realistic, intelligent, and adaptive practice scenarios. Need to practice handling objections from a hesitant seller? AI can push back just like a real client. Want to improve your cold-calling pitch or refine your listing presentation? AI can simulate different personalities and customer concerns to help you fine-tune your approach. Whether it's negotiating with a tough buyer, explaining commission rates, or mastering open house conversations, AI offers unlimited opportunities for targeted, on-demand training.

Beyond the scripts in this chapter, real estate agents can customize AI roleplaying to fit their unique challenges. To generate new role-playing ideas, agents should reflect on past client conversations that didn't go as planned or situations where they felt uncertain. Ask yourself:

- What objections do I frequently hear from buyers and sellers?
- Where do I feel the most nervous in conversations?
- What parts of my sales pitch need more clarity or confidence?

By identifying personal areas for improvement and instructing AI to simulate those scenarios, agents can sharpen their skills in a way that directly impacts their business.

In this chapter, we'll walk through specific AI-powered roleplaying exercises that can help you improve your sales scripts, handle objections with confidence, and elevate your client interactions. By leveraging AI as your personal sales coach, you can dramatically accelerate your learning curve, boost conversion rates, and ultimately close more deals.

For New Agents: Fundamental Skills

Introduction and Elevator Pitch Practice

Objective:

To confidently introduce yourself as a real estate agent in a way that is engaging, professional, and memorable. This script will help you practice introducing yourself in networking events, social settings, and client conversations.

Step 1: Set Up the Scenario

Before starting your roleplay with AI, decide the context:

- Are you introducing yourself at a networking event?
- Meeting a potential buyer or seller?
- Running into someone in a casual setting (like a coffee shop or gym)?

You can ask AI to play the role of a potential client, investor, or referral partner and respond as if they are meeting you for the first time.

Step 2: Example Script (Base Version)

You: *"Hi, I'm [Your Name], a local real estate agent with [Your Brokerage]. I specialize in helping [first-time buyers, investors, luxury clients—tailor it to your niche] find the perfect home. Whether you're thinking about buying, selling, or just curious about the market, I'd love to be a resource for you. What's your real estate goal right now?"*

Step 3: Enhancing the Script Based on AI Responses

The AI may respond in different ways. Here are some common responses and how to adjust your reply.

Scenario 1: The "I'm Not Looking Right Now" Response

AI: *"Thanks, but I'm not looking to buy or sell anytime soon."*
You: *"Totally understand! Most of my clients start by just getting market insights. I'd be happy to send you a quick update on what's happening in your area—would that be helpful?"*

Scenario 2: The "I Already Have an Agent" Response

AI: *"I'm actually working with another agent."*
You: *"That's great! Having a trusted agent is important. If anything ever changes or if things don't work out with your current agent, I'd be happy to help in any way I can."*

Scenario 3: The "Tell Me More" Response

AI: *"Actually, I've been thinking about buying but don't know where to start."*
You: *"That's exciting! I specialize in guiding buyers through the process. The first step is usually a quick chat to discuss your goals—would you like to set up a time to go over your options?"*

Step 4: Customizing and Practicing with AI

- **Adjust the script** based on your personality and niche (luxury, investors, first-time buyers, etc.).
- **Ask AI to challenge you**—request different objections to refine your responses.
- **Record your practice sessions** to review and improve delivery.

Final Thoughts:

First impressions matter, and a well-practiced elevator pitch can boost your confidence, generate leads, and make networking easier. By using AI to simulate real-world conversations, you can refine your introduction until it feels natural and effortless.

Calling a New Lead for the First Time

Objective:

To confidently make first contact with a new lead, introduce yourself professionally, build rapport, and set an appointment for a meeting or consultation. This script will help you practice making cold calls, following up on online leads, and reaching out to referrals.

Step 1: Set Up the Scenario

Before practicing with AI, decide the lead type:

- A **buyer lead** who registered on your website?
- A **seller lead** who requested a home valuation?
- A **referral** from a past client?

You can instruct AI to play the role of a buyer, seller, or potential investor with varying levels of interest.

Step 2: Example Script (Base Version)

You: *"Hi, is this [Lead's Name]? This is [Your Name] with [Your Brokerage]. I saw that you [signed up for home listings/requested a home valuation], and I wanted to personally follow up. Do you have a quick minute?"*

If they say *yes*, continue:

You: *"Great! I'd love to learn more about what you're looking for. Are you currently in the market, or just starting to explore your options?"*

Step 3: Enhancing the Script Based on AI Responses

Scenario 1: The "I'm Just Looking" Response

AI: *"I'm just browsing for now, not really serious."*
You: *"That's totally fine! Many of my clients start by just exploring the market. I can send you a quick market update with homes that match what you're looking for—would that be helpful?"*

Scenario 2: The "I'm Working with Another Agent" Response

AI: *"I already have an agent."*
You: *"That's great! Having an agent you trust is important. If anything ever changes or if things don't work out with your current agent, I'd be happy to help in any way I can."*

Scenario 3: The "Tell Me More" Response

AI: *"Yeah, I was looking at homes in [area], but I'm not sure what I qualify for."*
You: *"I can definitely help with that! The first step is usually getting pre-approved so we can look at homes in the right price range. Would you like me to connect you some great lenders, or should we set up a time to go over options?"*

Step 4: Customizing and Practicing with AI

- **Adjust the script** based on different lead sources (cold leads, warm referrals, online sign-ups).
- **Ask AI to challenge you**—request different objections so you can refine your responses.
- **Practice different tones**—friendly, professional, urgent—to see what works best.

Final Thoughts:

Your first conversation with a lead sets the tone for your relationship. AI roleplaying helps you refine your approach, handle objections confidently, and practice until setting appointments feels natural.

Handling the "I'm Just Looking" Objection

Objective:

To confidently handle the "I'm just looking" objection from potential buyers or sellers who are hesitant to take the next step. This script will help you turn casual inquiries into real opportunities while respecting the client's timeline.

Step 1: Set Up the Scenario

Before practicing with AI, decide the situation:

- A **buyer lead** browsing online listings.
- A **walk-in visitor** at an open house.
- A **seller lead** who is curious about home values but hesitant to list.

You can instruct AI to play the role of a hesitant client and provide different levels of resistance.

Step 2: Example Script (Base Version)

You: *"Hi [Lead's Name], I noticed you were looking at homes in [neighborhood/price range]. Are you thinking about making a move soon, or just browsing for now?"*
AI: *"I'm just looking for now."*
You: *"That's great! A lot of my clients start that way. Just out of curiosity, what's prompting you to check out homes in [area]?"*

Step 3: Enhancing the Script Based on AI Responses

Scenario 1: The "I'm Not Ready Yet" Response

AI: *"I won't be ready to buy for at least a year."*
You: *"Totally understandable! The market can change a lot in a year. Would you like me to send you occasional updates on home prices so you're well-informed when the time comes?"*

Scenario 2: The "I Don't Want to Talk to an Agent Yet" Response

AI: *"I'm just looking and don't want to commit to anything right now."*
You: *"No problem at all! My job isn't to pressure you but to be a resource when you're ready. If you ever have questions about neighborhoods, prices, or the buying process, feel free to reach out. Would you like access to my home search app so you can browse at your own pace?"*

Scenario 3: The "I'm Just Curious" Response

AI: *"I'm just looking at homes for fun, no real plans yet."*
You: *"That's great! Looking at homes is exciting. Many of my clients start by just browsing, then find something they love unexpectedly. If you ever want an inside scoop on upcoming listings or market trends, I'd be happy to keep you in the loop!"*

Step 4: Customizing and Practicing with AI

- **Modify the script** based on lead type (online lead, open house visitor, social media inquiry).
- **Ask AI to push back** with more resistance so you can refine your approach.

- **Try different closing techniques**—offering a home search app, sending market updates, or scheduling a casual consultation.

Final Thoughts:

Most buyers and sellers start as "just looking." The key is to keep the conversation open, offer value, and build trust so that when they're ready, you're the agent they call. AI roleplaying allows you to practice handling hesitation confidently and convert more leads over time.

Explaining Your Value as a Real Estate Agent

Objective:

To confidently articulate your value as a real estate agent, helping potential buyers and sellers understand why they should choose you over competitors or go the "For Sale By Owner" (FSBO) route. This script will help you establish credibility, differentiate yourself, and gain client trust.

Step 1: Set Up the Scenario

Before practicing with AI, decide who you're speaking to:

- A first-time homebuyer unsure about working with an agent.
- A seller considering a FSBO (For Sale By Owner) approach.
- A lead comparing multiple agents before making a decision.

You can instruct AI to act as a skeptical buyer or seller and challenge your value.

Step 2: Example Script (Base Version)

You: *"I completely understand that choosing the right real estate agent is a big decision. My job is to make your home buying or selling process as smooth and profitable as possible. I take pride in providing [market expertise, strong negotiation skills, and a data-driven approach] to ensure you get the best possible outcome. What's most important to you in an agent?"*

Step 3: Enhancing the Script Based on AI Responses

Scenario 1: The "I Can Do It Myself (FSBO)" Response

AI: *"I'm thinking about selling my home on my own to save money on commission."*
You: *"I totally understand! Many sellers consider FSBO to save on commission, but studies show that homes listed with an agent sell for 10-15% more on average. Not only do I help you price it right, but I also market it aggressively, negotiate on your behalf, and handle all the legal paperwork to protect your interests. Would you be open to a quick meeting to go over the numbers?"*

Scenario 2: "What Makes You Different?" Response

AI: *"I'm talking to a few different agents, why should I choose you?"*
You: *"Great question! What sets me apart is [your unique strength—maybe a deep understanding of the market, strong marketing skills, or AI-powered tools for data-driven pricing]. I also believe in [a high level of communication, a hands-on approach, or a proven track record of success]. More importantly, I tailor my strategy to fit your specific needs. What are you looking for in an agent?"*

Scenario 3: The "I'm Not Sure If I Need an Agent" Response

AI: *"I'm just starting the process and not sure if I need a real estate agent yet."*
You: *"That makes sense! Buying or selling a home is a big decision, and having expert guidance can make all the difference. I offer a no-obligation consultation where we go over your goals and how I can help, even if you're not ready to move forward yet. Would you be open to a quick call to explore your options?"*

Step 4: Customizing and Practicing with AI

- **Personalize the script** by adding your unique strengths and success stories.
- **Ask AI to push back** with stronger objections so you can refine your responses.
- **Practice delivering your answer** confidently and naturally.

Final Thoughts:

Being able to clearly communicate your value as a real estate agent is crucial for gaining clients. AI roleplaying allows you to practice different responses, refine your messaging, and build confidence so that when the opportunity arises, you're ready to win the client's trust.

Setting an Appointment for a Listing Presentation

Objective:

To confidently schedule a listing appointment with a potential seller by addressing their concerns, demonstrating value, and overcoming hesitation. This script will help you convert seller leads into face-to-face meetings, increasing your chances of winning the listing.

Step 1: Set Up the Scenario

Before practicing with AI, decide the seller's situation:

- A homeowner requesting a free home valuation.
- A seller unsure about listing now or waiting.
- A lead comparing multiple agents before choosing one.

You can instruct AI to play the role of a hesitant seller and provide different objections.

Step 2: Example Script (Base Version)

You: *"Hi [Seller's Name], this is [Your Name] with [Your Brokerage]. I saw you were interested in a home valuation, and I'd love to provide a detailed market analysis to show you what your home could sell for in today's market. When would be a good time for us to meet—this week or next?"*

Step 3: Enhancing the Script Based on AI Responses

Scenario 1: The "I'm Not Ready Yet" Response

AI: *"I'm not sure if I want to sell right now. I'm just curious about my home's value."*
You: *"That makes sense! Even if you're not ready yet, knowing your home's value can help you plan for the future. I can put together a free, in-depth market analysis, and we can discuss what the numbers look like—no pressure at all. Would Tuesday or Thursday work better for a quick meeting?"*

Scenario 2: The "I'm Talking to Other Agents" Response

AI: *"I'm speaking with a few different real estate agents before making a decision."*
You: *"That's a great approach! Every agent has a different strategy, and I'd love the opportunity to show you how I can help you get the best price in the shortest time. I use [your unique strategy—AI-driven pricing, advanced marketing, etc.] to make sure my clients get top dollar. How about a quick 20-minute chat so you can compare your options?"*

Scenario 3: The "Can You Just Send Me the Info?" Response

AI: *"Can you just email me the information instead?"*
You: *"Absolutely! I can send over some data, but to give you the most accurate estimate, I'd love to see your home in person and go over any updates or features that might impact the value. That way, you'll get the most precise market analysis. Would Thursday at 4 PM work for a quick walkthrough?"*

Step 4: Customizing and Practicing with AI

- **Personalize the script** based on your listing strategy and market knowledge.

- **Ask AI to challenge you** with tougher objections to refine your approach.
- **Practice different tones**—casual, confident, and persuasive—to see what works best.

Final Thoughts:

The key to securing listing appointments is making it easy for the seller to say yes while demonstrating your value. AI roleplaying allows you to practice handling objections smoothly and fine-tune your approach so you can book more appointments and win more listings.

For Mid-Level Agents: Overcoming Objections

Handling Commission Objections from Sellers

Objective:

To confidently respond when sellers question your commission, compare rates with other agents, or ask for a discount. This script will help you demonstrate your value, justify your commission, and overcome pricing objections with confidence.

Step 1: Set Up the Scenario

Before practicing with AI, decide the seller's concern:

- A seller comparing agents and looking for the lowest commission.
- A FSBO (For Sale By Owner) seller who doesn't want to pay an agent.
- A seller who wants a discount because they think the home will "sell itself."

You can instruct AI to play the role of a skeptical seller and challenge you with different objections.

Step 2: Example Script (Base Version)

Seller: *"I really like you, but another agent said they'd do it for less. Can you match their commission?"*
You: *"I understand that commission is a big part of your decision, and I want to make sure you get the most value for your money. The real question is: Do you want to save a little on commission, or do you want to walk away with the most money possible from your home sale?"*

Step 3: Enhancing the Script Based on AI Responses

Scenario 1: The "Another Agent Charges Less" Response

AI: *"Agent X said they'd do it for 4%. Why should I pay you 6%?"*
You: *"That's a great question! Some agents discount their fees because they can't justify full service. My goal is to not just sell your home, but to get you the highest price possible. Here's what I include that discount agents don't: [list your services—professional photography, AI-driven marketing, staging, targeted advertising, strong negotiation skills]. If I can get you $20,000 more in the sale price, is saving 1-2% on commission still worth it?"*

Scenario 2: The "The House Will Sell Itself" Response

AI: *"The market is hot. My home will sell no matter what, so why should I pay full commission?"*
You: *"You're right—the market is strong, and homes are selling. But the question is, do you want to sell fast, or do you want to sell for top dollar? The difference between a home selling at asking price vs. over asking often comes down to marketing, negotiation, and strategy. I specialize in getting sellers the most money possible. Wouldn't you rather maximize your profit?"*

Scenario 3: The "Can You Cut Your Commission?" Response

AI: *"I'd love to work with you, but can you do 5% instead of 6%?"*
You: *"I appreciate you asking! Here's the thing—agents who discount their fees often have to cut back on marketing and exposure to make up for it. I don't cut corners. I invest in top-tier marketing, negotiation strategies, and AI-powered tools to ensure my sellers get the best price possible. At the end of the day, I don't cost you money—I make you more money. Would you be open to reviewing exactly what I offer before making a final decision?"*

Step 4: Customizing and Practicing with AI

- **Personalize the script** by adding your own marketing strategy and success stories.
- **Ask AI to push back** harder on commission objections so you can practice confidently defending your value.
- **Try different closing techniques**—positioning commission as an investment, explaining marketing ROI, or showing real case studies of past sellers.

Final Thoughts:

Sellers don't always understand that not all agents offer the same level of service. Your job is to educate them on the value you bring, not just the cost of your commission. AI roleplaying allows you to practice handling these objections smoothly so you can stand firm on your value and win more listings.

Defending Your CMA (Comparative Market Analysis)

Objective:

To confidently present a Comparative Market Analysis (CMA) and justify your recommended pricing strategy, especially when sellers believe their home is worth more or want to "test the market" with a higher price. This script will help you educate sellers, set realistic expectations, and prevent overpricing issues.

Step 1: Set Up the Scenario

Before practicing with AI, decide the seller's pricing concern:

- A seller who believes their home is worth significantly more than market value.
- A seller who wants to overprice "just to see if someone will pay it."
- A seller who questions why a neighbor's home sold for more.

You can instruct AI to play the role of a seller resistant to your pricing recommendation and challenge you with objections.

Step 2: Example Script (Base Version)

You: *"I've put together a detailed Comparative Market Analysis (CMA) that looks at homes similar to yours that have recently sold, are currently listed, and those that didn't sell. Based on this data, I recommend listing your home at $X to $Y, which positions us competitively in today's market."*

Step 3: Enhancing the Script Based on AI Responses

Scenario 1: The "I Want to Price It Higher" Response

AI: *"I think my home is worth more than that. Can we list it $50,000 higher?"*
You: *"I completely understand—you've put a lot of love into your home, and it's natural to want the highest possible price. However, homes that are priced too high often sit on the market longer and end up selling for less. The data shows that homes priced correctly sell faster and get stronger offers. If we price too high, we risk scaring off buyers. Would you be open to reviewing how overpricing has impacted other listings in the area?"*

Scenario 2: The "Let's Test the Market" Response

AI: *"Why don't we just list it higher? We can always lower the price later."*
You: *"I get why that seems like a good idea, but here's what actually happens: When a home sits*

on the market too long, buyers start wondering what's wrong with it. Even if we reduce the price later, it may have already lost its appeal. The strongest offers usually come within the first two weeks, so pricing it right from the start gives you the best chance of getting top dollar. Does that strategy make sense?"

Scenario 3: The "My Neighbor's House Sold for More" Response

AI: *"My neighbor's home sold for $X—why is my home worth less?"*
You: *"Great question! Every home is unique, and a few key factors can affect value, including size, condition, upgrades, and market timing. Your neighbor's home may have had features that increased its value, such as a recent remodel or a larger lot. In our CMA, I've broken down exactly how your home compares so we can price it strategically to attract the right buyers."*

Step 4: Customizing and Practicing with AI

- **Add real examples**—mention specific homes in the seller's neighborhood to make your explanation more relatable.
- **Ask AI to challenge you** with stronger pushback so you can refine your approach.
- **Practice handling multiple objections** in one conversation—many sellers will bring up multiple concerns.

Final Thoughts:

Pricing is one of the biggest challenges in selling a home, and it's your job to guide sellers toward a data-driven decision. AI roleplaying helps you practice handling pricing objections smoothly, so you can set realistic expectations and prevent overpricing mistakes.

Negotiating an Offer for a Buyer

Objective:

To confidently negotiate on behalf of a buyer, ensuring they get the best possible price and terms in a competitive market. This script will help you handle counteroffers, navigate bidding wars, and overcome seller pushback to close a deal in your client's favor.

Step 1: Set Up the Scenario

Before practicing with AI, decide the type of negotiation:

- A seller countering above your client's offer.
- A multiple-offer situation where your client needs to be competitive.
- A buyer is asking for seller concessions (closing costs, repairs, etc.).

You can instruct AI to play the role of a listing agent or seller, pushing back on your negotiation attempts.

Step 2: Example Script (Base Version)

You: *"Thanks for sending over the counteroffer at [$X]. My client is very interested in this home, but they have a few concerns regarding [price, closing costs, contingencies, etc.]. I'd love to see if we can find a win-win solution that works for both parties. Here's what we're thinking..."*

Step 3: Enhancing the Script Based on AI Responses

Scenario 1: The "Seller Won't Budge on Price" Response

AI: *"The seller is firm on the price. They won't go lower."*
You: *"I understand they want to maximize their return. However, our offer is based on market data and comparable sales. Would the seller be open to covering some closing costs instead to help bridge the gap?"*

Scenario 2: The "Multiple-Offer Situation" Response

AI: *"We have multiple offers. Does your buyer want to improve their offer?"*
You: *"We definitely want to make this work for both sides. What's most important to the seller—price, closing timeline, or terms? We may be able to adjust our offer to make it the most attractive option."*

Scenario 3: The "Seller Won't Cover Repairs" Response

AI: *"The seller won't make any repairs. It's being sold as-is."*
You: *"I understand they're selling as-is. However, my client is concerned about [specific issue]. Would the seller consider a price adjustment or a credit at closing to address this?"*

Step 4: Customizing and Practicing with AI

- **Adjust the script** based on different negotiation styles (collaborative vs. aggressive).
- **Ask AI to increase resistance** so you can practice handling tougher negotiations.
- **Try different closing techniques**—emphasizing flexibility, offering creative solutions, or leveraging market data.

Final Thoughts:

Strong negotiation skills set top agents apart. AI roleplaying allows you to practice handling counteroffers, responding to seller pushback, and getting the best deal for your buyer—all without the pressure of a live deal.

Advanced Agents: Negotiations

Handling Multiple Offers on a Listing

Objective:

To confidently navigate multiple offers, advise sellers on selecting the strongest one, and negotiate terms that maximize profit while minimizing risk. This script will help you balance price, contingencies, financing strength, and closing terms to make the best decision for your seller.

Step 1: Set Up the Scenario

Before practicing with AI, decide the type of multiple-offer situation:

- Two or more similar offers – which one should the seller choose?
- One offer is higher, but another has better terms (stronger financing, no contingencies, etc.).
- A buyer wants to revise their offer after learning about competing bids.

You can instruct AI to play the role of a seller debating which offer to accept or a buyer's agent pushing for their client's offer to be accepted.

Step 2: Example Script (Base Version)

You: *"We've received multiple offers on your home, which is a fantastic position to be in! I've put together a side-by-side comparison of each offer so we can evaluate them based on price, contingencies, financing, and closing terms. Let's go through these together and determine which one makes the most sense for your goals."*

Step 3: Enhancing the Script Based on AI Responses

Scenario 1: The "Highest Offer vs. Best Terms" Dilemma

AI (as seller): *"This offer is the highest price, but the other one has a stronger loan and no contingencies. Which should I take?"*
You: *"Great question! While the highest offer looks tempting, it's important to consider the risk. If the buyer's financing falls through or they back out due to contingencies, we'll be back to square one. A strong, clean offer with solid financing may be the safer bet. Would you like to counter the higher offer to remove some of the contingencies?"*

Scenario 2: The "Escalation Clause" Question

AI (as seller): *"One buyer included an escalation clause—what does that mean for us?"*
You: *"An escalation clause means this buyer is willing to beat any other offer up to a certain price. This could work in our favor if we want to push the price higher. We can either accept their max price or go back to the top two buyers and ask for their highest and best offers. Would you like to take that approach?"*

Scenario 3: The "Buyer Wants to Improve Their Offer" Response

AI (as buyer's agent): *"My client really wants this home. What can they do to make their offer stronger?"*
You: *"We're in a competitive situation, so your client may want to consider increasing their price, waiving contingencies, or offering a larger earnest money deposit. Is your buyer willing to make any of those adjustments?"*

Step 4: Customizing and Practicing with AI

- **Adjust the script** based on different seller personalities: some prioritize price, while others value a stress-free sale.
- **Ask AI to introduce challenges** like a financing issue or buyer hesitation to test your adaptability.
- **Practice different negotiation techniques**—asking for "highest and best," leveraging escalation clauses, or countering strategically.

Final Thoughts:

Handling multiple offers requires a mix of strategy, negotiation, and risk management. AI roleplaying helps you practice advising sellers, responding to competing offers, and ensuring your client walks away with the best possible deal.

Navigating a Difficult Inspection Negotiation

Objective:

To confidently handle tough inspection negotiations, keep buyers and sellers on track, and prevent deals from falling apart due to repair requests or inspection concerns. This script will help you find solutions that satisfy both parties while protecting your client's best interests.

Step 1: Set Up the Scenario

Before practicing with AI, decide the type of inspection challenge:

- The buyer requests excessive repairs.

- The seller refuses to make any repairs.
- Major issues (foundation, roof, plumbing, etc.) arise, threatening the deal.

You can instruct AI to play the role of a buyer's agent requesting repairs or a seller resisting concessions.

Step 2: Example Script (Base Version)

You (as listing agent): *"I understand the inspection revealed some concerns, and we want to find a fair solution that keeps the deal moving forward. Let's discuss what's reasonable and how we can work together to get this resolved."*

Step 3: Enhancing the Script Based on AI Responses

Scenario 1: The "Buyer Wants Everything Fixed" Response

AI (as buyer's agent): *"My client wants all 15 items on the inspection report fixed before moving forward."*
You: *"I completely understand they want a home in good condition. However, some of these items are general maintenance rather than major defects. Let's prioritize the most critical repairs that could impact financing or safety. Would your buyer be open to focusing on the key issues instead?"*

Scenario 2: The "Seller Refuses to Do Any Repairs" Response

AI (as seller): *"I'm not fixing anything. They can take it or leave it."*
You: *"I understand that making repairs wasn't part of your plan, but we're close to the finish line. If we don't offer anything, we risk losing this buyer and having to start over. Would you consider offering a small credit at closing to help move the deal forward?"*

Scenario 3: Major Issue Found in Inspection

AI (as buyer's agent): *"The inspection found a major roof problem. My client is considering walking away."*
You: *"I completely understand—that's a big concern. Would your client be open to getting a second opinion from a licensed contractor? We may be able to negotiate a repair or seller credit instead of canceling the deal."*

Step 4: Customizing and Practicing with AI

- **Adjust the script** based on different property conditions (older homes vs. new builds).
- **Ask AI to introduce stronger resistance** to test your problem-solving skills.

- **Practice different resolution strategies**—offering seller credits, negotiating repairs, or bringing in experts for further evaluation.

Final Thoughts:

Inspection negotiations can be one of the most stressful parts of a transaction. AI roleplaying helps you practice finding creative solutions, keeping both parties calm, and ensuring the deal moves forward smoothly.

Chapter 7 Summary

Mastering the art of communication is one of the most valuable skills a real estate agent can develop, and roleplaying is the fastest way to refine those skills. Whether it's handling objections, negotiating deals, or confidently presenting your value, practicing conversations in a risk-free environment can make all the difference. Traditionally, agents roleplay with colleagues or mentors, but now, AI-powered roleplaying tools offer 24/7 access to unlimited practice scenarios—without fear of judgment or embarrassment.

This chapter explored several key real estate scripts that agents can practice using AI, covering a wide range of situations from new agent fundamentals (introducing yourself, handling objections, setting appointments) to advanced negotiation tactics (multiple-offer situations, commission objections, and inspection disputes). By engaging in AI roleplaying, agents can refine their ability to think on their feet, improve confidence, and deliver smoother, more persuasive conversations when interacting with real clients.

Whether it's perfecting a listing presentation, learning to defuse tough client objections, or strengthening negotiation tactics, AI-powered roleplaying is a game-changing tool that can take a real estate agent from good to great. With consistent practice, agents will close more deals, build stronger client relationships, and ultimately elevate their careers.

8. FUTURE AI STRATEGIES

Leveraging Advanced AI Strategies in Real Estate

AI is revolutionizing real estate, offering cutting-edge tools that can help agents generate more leads, market more effectively, and streamline transactions. The strategies in this chapter go beyond basic automation and dive into high-level AI applications that can give real estate agents a serious competitive edge. From predictive analytics to smart contracts and AI-powered virtual tours, these advanced techniques are designed to enhance—not replace—the skills of an experienced agent.

Author Note: Most of these DO NOT WORK today, but will likely be available in the near future. This chapter is here to give you a sense where the industry is going.

It's important to note that some of these AI tools may require paid subscriptions, and their effectiveness will vary depending on market conditions, brokerage policies, and individual business models. As with any new technology, always consult with your managing broker before implementing advanced AI strategies, especially when using AI in contracts, negotiations, or client interactions.

While AI can automate tasks and provide powerful insights, it cannot replace the expertise, negotiation skills, and human touch that great real estate agents bring to the table. These strategies are not a substitute for mastering the fundamentals of real estate—they are a supplement that can help agents work smarter, scale their business, and stay ahead of the competition.

As AI continues to evolve, real estate agents who leverage cutting-edge AI strategies will have a significant competitive advantage. These advanced tactics go beyond basic automation and chatbots—these are high-level AI applications that can revolutionize lead generation, marketing, and client service. Whether you're looking to attract high-value leads, optimize your marketing, or increase efficiency in transactions, these ten advanced AI strategies can take your real estate business to the next level.

Ten Future Advanced AI Strategies

Lead Generation and Client Targeting

1. **AI-Driven Predictive Analytics for Finding Motivated Sellers**

    - Use AI-powered data platforms to predict which homeowners are most likely to sell based on factors like mortgage payments, home equity, life events, and market trends.
    - AI tools like Revaluate, Likely.ai, or SmartZip analyze thousands of data points to identify high-probability sellers before they even consider listing.

2. **AI-Powered Lead Scoring & Qualification**

    - Instead of wasting time on cold leads, AI-driven CRM systems (like Follow Up Boss or kvCORE) can rank and score leads based on their engagement, online activity, and likelihood to transact.
    - AI can analyze social media activity, property searches, and email interactions to predict which leads are most ready to buy or sell.

3. **Automated AI-Generated CMA Reports and Market Forecasting**

    - AI can create hyper-detailed Comparative Market Analyses (CMAs) instantly, factoring in market trends, neighborhood changes, and predictive pricing models.
    - Advanced AI-powered valuation models can predict future home prices and appreciation rates based on economic trends and local market conditions.

Marketing and Branding with AI

1. **AI-Generated Personalized Video Marketing**

    - AI tools like Synthesia or DeepBrain AI can create customized, hyper-personalized video messages for clients, using AI generated avatars that look and sound like you.
    - Use AI to generate personalized listing videos, market updates, and client follow-up messages—without manually recording every video.

2. **AI-Powered SEO** and Content Creation for Local Domination

    - AI can analyze search trends and local buyer/seller behavior to suggest high-ranking blog topics and YouTube video ideas.
    - Tools like ChatGPT and Jasper can create keyword-optimized blog posts, YouTube scripts, and social media content to dominate local searches.

3. **AI-Powered Social Media Advertising and Audience Targeting**

- AI tools like Adzooma and Albert AI can automatically run, optimize, and adjust digital ads based on performance.
- AI can analyze which types of posts generate the most engagement and automatically schedule content when your audience is most active.

Automation and Efficiency

1. **AI Chatbots That Learn and Adapt to Client Conversations**

- Instead of basic autoresponders, AI-powered chatbots (like ChatGPT API, Drift, or ManyChat) can learn from past conversations and provide natural, human-like responses to buyers and sellers.
- AI chatbots can automate appointment scheduling, answer complex real estate questions, and nurture leads 24/7.

2. **AI-Powered Smart Contracts and Transaction Management**

- AI tools can analyze contract clauses, compliance requirements, and legal risks in real estate transactions.
- Smart contracts (using blockchain & AI) automate the closing process, reducing paperwork, delays, and manual errors.

Client Service and Market Domination

1. **AI-Generated Custom Property Investment Reports**

- AI tools like Mashvisor and PropStream can generate detailed investment analysis reports showing rental income potential, cap rates, and ROI for investor clients.
- AI can customize reports for individual investors based on their portfolio goals, budget, and risk tolerance.

2. **AI-Powered Virtual Reality (VR) and Augmented Reality (AR) Home Showings**

- AI can enhance virtual home tours by allowing buyers to change paint colors, swap furniture, or see a remodeled version of a home in real-time.
- AI-driven AR apps can allow potential buyers to scan a property's QR code and see projected home values, tax info, or historical sale prices instantly.

9. PARTING THOUGHTS

The real estate industry is evolving, and AI is at the forefront of this transformation. Those who embrace AI early will gain a competitive advantage, improve efficiency, and provide better service to their clients. The key is to integrate AI tools strategically, balancing automation with the human touch that remains at the heart of real estate success.

Integrating advanced AI strategies into your real estate business can be a game-changer, but it requires time, investment, and a willingness to learn new skills. These tools are designed to enhance lead generation, marketing, negotiation, and transaction management, but they won't replace the essential skills of an experienced real estate agent. The key to success isn't trying to implement every AI strategy all at once but rather choosing one or two that align most directly with your current business goals. Whether it's using predictive analytics to identify motivated sellers, AI-driven marketing to attract high-quality leads, or smart contracts to streamline closings, the right approach can help you work smarter, scale faster, and improve client experiences.

It's important to remember that some AI-powered tools require financial investment, a learning curve, and technical setup. If you're new to AI, start small—identify your biggest business challenge and choose an AI solution that directly addresses that problem. For example, if you struggle with lead generation, predictive analytics and AI-powered social media ads could help you generate better leads. If you want to stand out with innovative marketing, tools like AI-generated video and automated content creation might be a smart first step. The goal is to integrate AI gradually, testing what works and refining your approach over time.

AI won't replace the personal connections and expertise that make a real estate agent truly successful, but it can provide data-driven insights, automation, and scalability to take your business to the next level. By embracing the right AI strategies for your business, you'll stay ahead of the competition, improve efficiency, and create a high-tech, high-touch client experience that builds long-term success.

By using AI for lead generation, marketing, and client communication, agents can scale their business while focusing on what matters most—building relationships and closing deals.

AI is not here to replace real estate agents—it's here to empower them. Those who integrate AI strategically will enhance efficiency, improve client service, and stay ahead in an evolving market.

10. GLOSSARY

AI for Real Estate Agents: Glossary of Key Terms
This glossary provides simple, jargon-free definitions of AI-related terms that real estate agents, managing brokers, and brokerage owners may encounter when integrating AI into their business.

A

- **AI (Artificial Intelligence)** – AI refers to computer programs that analyze data, make predictions, and automate tasks that typically require human thinking. In real estate, AI enhances marketing, lead generation, and client communication.
- **AI Chatbot** – An automated assistant that responds to messages on websites or social media, answering common questions, collecting client information, and scheduling appointments.
- **AI-Enhanced CRM (Customer Relationship Management)** – A system that tracks client interactions, automates follow-ups, and personalizes communication. AI-driven CRMs analyze behavior and suggest the best times to engage leads.
- **AI Model** – A type of software that uses data to learn patterns and make decisions, similar to human reasoning. Real estate agents use AI models to automate emails, predict home prices, and identify top leads.
- **AI-Powered Content Creation** – The use of artificial intelligence to generate written, visual, or video content for marketing. AI tools can write blog posts, create social media captions, and generate real estate listing descriptions.
- **AI-Powered CRM (Customer Relationship Management)** – A software system that tracks client interactions, organizes contacts, and automates follow-ups. AI-enhanced CRMs predict client needs and suggest the best next steps.
- **AI-Powered Event Planning** – AI tools that assist with organizing client appreciation events by automating invitations, scheduling, and post-event follow-ups.
- **AI-Powered Video Content** – Pre-recorded or AI-generated videos that answer common real estate questions, reducing the need for repetitive explanations. AI tools can generate scripts, add voiceovers, or even create personalized video responses.
- **AI-Generated Real Estate Marketing Emails** – AI-created emails personalized for leads and clients. AI suggests subject lines, writes the email body, and schedules follow-ups based on client interactions.
- **AI for Lead Scoring** – AI assigns a "score" to each lead based on their level of interest and engagement. High-scoring leads are more likely to convert into clients.

- **Automated Follow-Up** – AI systems that send pre-written emails or texts to stay in touch with clients and leads. These follow-ups keep agents top-of-mind without requiring manual effort.
- **Automated Video Content** – AI-created videos that answer client FAQs, reducing the need for manual explanations.

B

- **Blog Content Generation** – AI helps real estate agents write blog posts by suggesting topics, structuring content, and generating drafts.

C

- **Chatbot** – A computer program that can have conversations with people. In real estate, chatbots help answer client questions, schedule showings, and provide property details automatically.
- **Client Sentiment Analysis** – AI can analyze words and phrases in emails, reviews, or social media to understand how a client feels about a home or an agent's service.
- **Comparative Market Analysis (CMA)** – A report that compares a home to recently sold properties. AI improves accuracy by analyzing real-time pricing trends.
- **Content Scheduling** – AI allows real estate agents to plan and schedule social media posts at optimal times for engagement.

D

- **Data Privacy** – Protecting client information from being shared or misused. real estate agents using AI must follow best practices to keep personal and financial data secure.
- **Document Processing Automation** – AI tools that scan, analyze, and organize paperwork, reducing errors in contracts, disclosures, and other documents.
- **Document Summarization AI** – AI extracts key insights from lengthy real estate documents (e.g., HOA documents, inspection reports) and presents a concise summary.
- **Drip Campaigns** – A sequence of automated messages (emails, texts, or calls) sent over weeks or months to nurture leads. AI personalizes these campaigns based on client behavior.

F

- **Faceless Videos** – AI-generated videos using images, voiceovers, and text instead of on-camera appearances.

G

- **Generative AI** – AI that creates new content, such as property descriptions, listing photos, video scripts, and emails. Real estate agents use generative AI to automate content creation and marketing.
- **Generative Role-Playing AI** – AI simulates real-life buyer-agent conversations, helping real estate agents practice responses to objections, negotiations, and client engagement.
- **Google Business Profile** – A free tool from Google that helps real estate agents appear in local search results. AI optimizes these listings by adding the right keywords and responding to reviews.

I

- **Intelligent Email & Text Automation** – AI tools that analyze client interactions and send personalized, well-timed messages to keep buyers engaged.
- **Iterative Prompting** – The process of refining and improving an AI prompt to get a better response. Instead of asking once, agents can tweak their request multiple times to get a more useful answer.

K

- **Keyword Optimization** – AI suggests the best words and phrases for online content to improve search engine visibility and drive traffic to a real estate agent's website.

L

- **Lead Generation** – The process of finding potential buyers and sellers. AI helps identify high-quality leads by analyzing online behavior and predicting who might be ready to move.
- **Lead Nurturing** – Building relationships with potential clients over time by providing useful information, answering questions, and staying in touch. AI automates this with personalized emails and reminders.
- **Lead Qualification** – AI analyzes engagement data (such as email opens or website visits) to rank leads as "hot" or "cold."
- **Lead Scoring** – AI ranks leads based on how likely they are to buy or sell soon, helping real estate agents prioritize follow-ups.

M

- **Machine Learning** – A branch of AI that enables computers to learn from past data and improve over time without being explicitly programmed.
- **Market Analysis** – AI speeds up the process of examining real estate trends, pricing, and supply-and-demand factors by compiling and summarizing vast amounts of data.

- **Market Trend Analysis** – AI reviews housing market data, such as price changes and home sales speed, to help real estate agents and clients make informed decisions.
- **Meta Descriptions** – Short text summaries that appear in search results when someone looks up a website. AI can generate meta descriptions to improve search engine visibility.

P

- **Personalized Marketing** – AI tailors ads, emails, and property recommendations based on a client's interests and past searches.
- **Predictive Analytics** – AI predicts future trends, such as which homeowners might sell soon or which neighborhoods will increase in value.
- **Predictive Buyer Insights** – AI analyzes a buyer's online activity, search history, and engagement patterns to predict when they are ready to make an offer.
- **Property Listings Automation** – AI creates and updates online property listings with descriptions, photos, and pricing recommendations.

R

- **Referral Marketing** – Encouraging past clients to refer new business to an agent. AI automates follow-ups and reminds agents to request reviews and testimonials.
- **Reputation Management** – AI tools monitor online reviews and social media mentions, suggesting professional responses to maintain a positive public image.
- **Review Automation** – AI-powered tools that request, collect, and summarize client reviews for marketing purposes.

S

- **SEO (Search Engine Optimization)** – AI improves website ranking on Google by suggesting keywords, optimizing content, and analyzing competitors' strategies.
- **Smart CRM** – AI-powered systems that track client interactions, suggest follow-ups, and help agents manage leads efficiently.
- **Social Proof** – Positive client reviews, testimonials, and engagement that show a real estate agent is trustworthy. AI helps collect and showcase these reviews.
- **Speech-to-Text AI** – AI transcription tools that convert spoken conversations into text, allowing real estate agents to quickly summarize meetings and follow up with clients.
- **Staging AI** – AI-generated images or virtual furniture placement that helps buyers visualize a home's potential.

T

- **Text-to-Speech (TTS)** – AI technology that converts written text into spoken words, useful for voiceovers in real estate videos.
- **Transaction Automation** – AI streamlines real estate paperwork, deadlines, and administrative tasks.
- **Transaction Coordination** – AI-powered tools that automate contract management, deadline tracking, and compliance tasks.

V

- **Video Script AI** – AI tools that generate structured video scripts for real estate marketing and client education.
- **Virtual Assistants** – AI-powered tools that handle routine tasks like responding to messages, scheduling appointments, and following up with leads.
- **Voice Cloning** – AI technology that replicates a person's voice, allowing real estate agents to generate marketing messages or voicemail greetings without recording new audio.

APPENDIX A – List of Buyer Videos

Buyer Onboarding and Preparation

Video 1 – New Buyer Client Onboarding Script

Video 2 – Preparing for a Buyer Consultation

Video 3 – Why Get Pre-Qualified for a Mortgage

Video 4 – Understanding Buyer Closing Costs

Video 5 – Your Dream First Home vs. What You Can Afford

The Home Search and Making an Offer

Video 6 – Why Buying is Better Than Renting for Most

Video 7 – Elements of the Home Search

Video 8 – What It's Like to Go Under Contract

Video 9 – Common Contract Pitfalls and How to Avoid Them

Video 10 – Negotiating the Contract

Video 11 – How to Handle Multiple Offers

Video 12 – What Happens If the Deal Falls Apart?

Due Diligence and Financing

Video 13 – What is a Home Inspection? Questions to Ask

Video 14 – Reviewing the Title Policy: Questions to Ask

Video 15 – Reviewing HOA Documents: Questions to Ask

Video 16 – Understanding Home Warranties

Video 17 – Real Estate Tax Benefits and Consideration

Video 18 – Understanding the Home Appraisal Process

Video 19 – What Happens If the Appraisal Comes in Low?

Preparing for Closing and Homeownership

Video 20 – Credit Score Best Practices While Under Contract

Video 21 – Life Preparation While Under Contract

Video 22 – Preparing for Closing: Settlement Statement

Video 23 – Going to Closing: What to Expect and Best Practices

Video 24 – After Closing – Now What?

Video 25 – Building Wealth Through Homeownership (Why Your First Home is Just the Beginning of Your Real Estate Journey)

Visit the Buyer Video Library

APPENDIX B – List of Seller Videos

Seller Onboarding and Preparation

Video 1 – Understanding the Home Selling Process

Video 2 – How to Choose the Right Listing Agent

Video 3 – Pre-Listing Home Inspection: Is It Worth It?

Video 4 – The Top Home Repairs to Maximize Value

Video 5 – Staging Your Home for Maximum Buyer Appeal

Pricing and Marketing Strategy

Video 6 – How to Price Your Home Right the First Time

Video 7 – The Power of Comparative Market Analysis (CMA)

Video 8 – Common Pricing Mistakes and How to Avoid Them

Marketing and Listing Exposure

Video 10 – How Professional Photography and Videos Help Sell Faster

Video 11 – Creating an Engaging Online Listing That Stands Out

Video 12 – The Role of Social Media in Selling Your Home

Video 13 – Understanding MLS and Syndication for Exposure

Showings and Buyer Interaction

Video 14 – How to Prepare for Showings and Open Houses

Video 15 – What Buyer Feedback Tells You and How to Adjust

Video 16 – How AI Helps Optimize Showing Schedules

Offers and Negotiation

Video 17 – Understanding the Offer Process for Sellers

Video 18 – How to Negotiate the Best Deal for Your Home

Video 19 – What Happens During a Home Inspection

Video 20 – The Appraisal Process and What It Means for Your Sale

Closing and Moving Forward

Video 21 – What to Expect After Going Under Contract

Video 22 – Navigating the Closing Process Like a Pro

Video 23 – Common Closing Delays and How to Prevent Them

Video 24 – Moving Out: Timeline, Tips, and Final Responsibilities

Video 25 – Life After Selling: What's Next?

Visit the Seller Video Library

Unlock Your Clients' Path to Financial Freedom – Get *The Mortgage Escape Plan*!

If you found value in *How Real Estate Agents Can Succeed with AI*, then you already understand the power of working smarter, not harder. But what if you could help your clients do the same with their mortgage—**paying it off in 7-10 years instead of 30?**

That's exactly what *The Mortgage Escape Plan* is all about! This game-changing resource gives real estate agents like you the tools to educate your buyers and homeowners on how to maximize their mortgage strategy, build long-term wealth, and break free from unnecessary debt faster than they ever thought possible.

Why You Need This Book:

- **Empower Your Clients** – Be the trusted advisor who helps them not just buy a home, but create financial freedom.
- **Differentiate Yourself** – Stand out from other agents by offering real financial solutions that make a lasting impact.
- **Boost Repeat & Referral Business** – When clients see how you've helped them save hundreds of thousands in interest, guess who they'll recommend? (Hint: YOU!)
- **Strengthen Your Own Wealth Strategy** – These principles aren't just for clients—you can apply them to your own real estate investments, too!

Get Your Free Copy Today!

This book is already helping thousands of homeowners rethink their mortgage strategy and take control of their financial future. Now it's your turn! Download *The Mortgage Escape Plan* for free at: www.MasseyMoneyEducation.com/Mortgage-Escape-Ebook

Give yourself (and your clients) the ultimate advantage—because owning a home is great, but owning it debt-free is even better!

Elevate Your Real Estate Career with Exclusive Opportunities from Your Castle Real Estate!

As a dedicated real estate professional, you're always seeking ways to **enhance your skills**, **strategize effectively**, and **stay ahead** in a competitive market. At **Your Castle Real Estate**, we share your commitment to excellence and offer **exclusive opportunities** to support your growth:

1. Attend Our Cutting-Edge Training Classes

Stay at the forefront of industry trends and technological advancements with our comprehensive training programs. Our classes cover a wide range of topics, including:

- **Advanced Marketing Techniques**: Learn innovative strategies to attract and retain clients.
- **AI Integration in Real Estate**: Discover how artificial intelligence can revolutionize your business operations.
- **Negotiation Mastery**: Enhance your skills to close deals more effectively.

These sessions are designed to provide actionable insights that you can implement immediately.

2. Join Our Strategic Business Planning Sessions

Chart a clear path to success with our personalized business planning workshops. Collaborate with industry experts to:

- **Set Achievable Goals**: Define clear objectives for your business growth.
- **Develop Customized Strategies**: Create tailored plans that align with your unique vision.
- **Monitor Progress**: Establish metrics to track your achievements and adjust strategies as needed.

These sessions empower you to take control of your career trajectory with confidence.

3. Participate in Our Quarterly Market Insight Meetings

Gain a competitive edge with our in-depth market analysis meetings held every quarter. In these gatherings, we:

- **Analyze Current Market Trends**: Understand the latest developments affecting the real estate landscape.
- **Share Proprietary Data**: Access exclusive information to inform your business decisions.
- **Foster Networking**: Connect with like-minded professionals to share experiences and strategies.

Staying informed is key to anticipating changes and seizing opportunities.

Why Partner with Your Castle Real Estate?

With over 20 years of experience and a deep understanding of the Denver real estate market, Your Castle Real Estate is committed to empowering agents like you. Our resources are designed to provide you with the knowledge and tools necessary to excel.

Take the Next Step in Your Career

Ready to elevate your real estate business? Visit www.yourcastle.com to learn more about our upcoming events and register for the opportunities that align with your goals.

Invest in yourself today and watch your real estate career soar to new heights!

Made in United States
Orlando, FL
11 November 2026

90609410R00115